SpringerBriefs in Speech technology

More information about this series at http://www.springer.com/series/10059

Hamidreza Chinaei • Brahim Chaib-draa

# Building Dialogue POMDPs from Expert Dialogues

## An end-to-end approach

Springer

Hamidreza Chinaei
University of Toronto
Toronto, ON, Canada

Brahim Chaib-draa
Université Laval
Quebec, QC, Canada

ISSN 2191-8112 ISSN 2191-8120 (electronic)
SpringerBriefs in Speech technology
ISBN 978-3-319-26198-0 ISBN 978-3-319-26200-0 (eBook)
DOI 10.1007/978-3-319-26200-0

Library of Congress Control Number: 2015954569

Springer Cham Heidelberg New York Dordrecht London


Printed on acid-free paper

Springer International Publishing AG Switzerland is part of Springer Science+Business Media (www.springer.com)

# Contents

# Acronyms

The following basic notation is used in this book:

| | |
|---|---|
| $\boldsymbol{x}$ | Bold lower-case letters represent vectors |
| $\boldsymbol{X}$ | Bold upper-case letters represent matrices |
| $a$ | Italic letters refer to scalar values |
| $x \leftarrow a$ | Assignment of $x$ to value of $a$ |
| $\Pr(s)$ | The discrete probability of event $s$, the probability mass function |
| $p(x)$ | The probability density function for a continuous variable $x$ |
| $\{a_1, \ldots, a_n\}$ | A finite set defined by the elements composing the set |
| $N$ | Number of intents |
| $K$ | Number of features |
| $B$ | Number of trajectories |

Commonly used acronyms include:

| | |
|---|---|
| ASR | Automatic speech recognition |
| IRL | Inverse reinforcement learning |
| MDP | Markov decision process |
| PBVI | Point-based value iteration |
| POMDP | Partially observable Markov decision process |
| RL | Reinforcement learning |
| SDS | Spoken dialogue systems |

Acronyms defined in this book include:

| | |
|---|---|
| MDP-IRL | IRL in the MDP framework |
| POMDP-IRL | IRL in the POMDP framework |
| PB-POMDP-IRL | Point-based POMDP-IRL |
| POMDP-IRL-BT | POMDP-IRL using belief transition estimation |
| POMDP-IRL-MC | POMDP-IRL using Monte Carlo estimation |

# Chapter 1
# Introduction

Spoken dialog systems (SDSs) are the systems that help the human user to accomplish a task using the spoken language. For example, users can use an SDS to get information about bus schedules over the phone or Internet, to get information about a tourist town, to command a wheelchair to navigate in an environment, to control a music player in an automobile, to get information from customer care to troubleshoot devices, and many other tasks. Building SDSs is a difficult problem since automatic speech recognition (ASR) and natural language understanding (NLU) make errors which are the sources of uncertainty in SDSs. In addition, the human user behavior is not completely predictable. The users may change their *intents* during the dialog, which makes the SDS environment even more uncertain.

Consider the example in Table 1.1 taken from SACTI-2 data set of dialogs (Weilhammer et al. 2004), where SACTI stands for simulated ASR-channel: tourist information. The first line of the table shows the user utterance, $u_1$. Because of the ASR errors, this utterance is recognized by the system as the line in the braces, i.e., $\tilde{u}_1$. The next line, $m_1$, shows the system's (machine's) response to the user.

For each dialog utterance, the system's goal is first to capture the user intent and then to execute the best action that satisfies the user intent. For instance, consider the received user utterance in the first dialog turn: $\tilde{u}_1$ : [Is there a good restaurant week an hour tonight]. In this turn, the system can predict the user intent as information request for *food* places since the utterance contains the only keyword *restaurant*. However, it is not the case for the received user utterance in the second turn: $\tilde{u}_2$ : [No I think late like uh museum price restaurant]. This utterance contains misleading words such as *museum* that can be a strong observation for other user intents, such as user intent for *visiting* areas. Ideally, in $\tilde{u}_2$ the system should estimate that the user is looking for *food* places, and consider *museum* as a corrupted word, since the user has been asking for *restaurant* in his previous utterance, $\tilde{u}_1$. It may however consider *museum* as a change of the user intent, and mis-estimate that the user is looking for *visiting* areas.



H. Chinaei, B. Chaib-draa, *Building Dialogue POMDPs from Expert Dialogues*, SpringerBriefs in Speech technology, DOI 10.1007/978-3-319-26200-0_1

**Table 1.1** A sample from the SACTI-2 dialogs (Weilhammer et al. 2004)

| | |
|---|---|
| $u_1$ : | Is there a good restaurant we can go to tonight |
| $\tilde{u}_1$ : | [Is there a good restaurant week an hour tonight] |
| $m_1$ : | Would you like an expensive restaurant |
| | |
| $u_2$ : | No I think we'd like a medium priced restaurant |
| $\tilde{u}_2$ : | [No I think late like uh museum price restaurant] |
| $m_2$ : | Cheapest restaurant is eight pounds per person |
| | |
| $u_3$ : | Can you tell me the name |
| $\tilde{u}_3$ : | [Can you tell me the name] |
| $m_3$ : | bochka |
| | b o c h k a |
| | |
| $u_4$ : | Thank you can you show me on the map where it is |
| $\tilde{u}_4$ : | [Thank you can you show me i'm there now where it is] |
| $m_4$ : | It's here |
| | |
| $u_5$ : | Thank you |
| $\tilde{u}_5$ : | [Thank u] |
| $u_6$ : | I would like to go to the museum first |
| $\tilde{u}_6$ : | [I would like a hour there museum first] |
| | ... |

Similarly, the system has to resolve another problem in the received utterance in the third turn: $\tilde{u}_3$ : [Can you tell me the name]. Here, there is no keyword *restaurant*, however, the system needs to estimate that the user is actually requesting information for *food* places basically because the user has been asking about *food* places in the previous utterances.

In addition, the NLU is challenging. For instance, there are several ways of expressing an *intent*. This is notable, for instance, in SmartWheeler, which is an intelligent wheelchair to help persons with disabilities (Pineau et al. 2011). SmartWheeler is equipped with an SDS, thus the users can give their commands through the spoken language besides a joystick. The users may say a command in different ways. For instance for turning right, the user may say:

- *turn right a little please*,
- *turn right*,
- *right a little*,
- *right*.

And many other ways to say the same intent. As a response, SmartWheeler can perform the TURN RIGHT A LITTLE action or ask for REPEAT.

Such problems become more challenging when the user utterance is corrupted by ASR. For instance, SmartWheeler may need to estimate that the user asks for *turn right* from the ASR output, *10 writer little*. We call domains such as SmartWheeler *intent-based* dialog domains. In such domains, the user intent is the dialog *state* which should be estimated by the system to be able to perform the best action.

In this context, performing the best action in each dialog state (or the estimated dialog state) is a challenging task due to the uncertainty introduced by ASR errors and NLU problems as well as the uncertain environment as a result of change in the user behavior. In such uncertain domains where the decision making is sequential, the suitable formal framework is the Markov decision process (MDP). However, the MDP framework considers the environment as fully observable and this does not conform to real applications which are partially observable such as SDSs. In this context, the partially observable MDP (POMDP) framework can deal this constraint of uncertainty.

In fact, the POMDP framework has been used to model the uncertainty of SDSs in a principled way (Roy et al. 2000; Zhang et al. 2001a,b; Williams and Young 2007; Thomson 2009; Gašić 2011; Young et al. 2013). The POMDP framework is an optimization framework that supports automated policy solving by optimizing a *reward function*, while considers the states partially observable. In this framework, the reward function is the crucial component that directly affects the optimized policy and is a major topic of this book, and is discussed below in this chapter. The optimized policy depends also on other components of the POMDP framework. The POMDP framework includes model components such as: a set of states, a set of actions, a set of observations, a transition model, an observation model, a reward function, etc.

For the example shown in Table 1.1, if we model the control module as a dialog POMDP, the POMDP states can be considered as the possible user intents (Roy et al. 2000), i.e., the user information need for *food* places, *visit* areas, etc. The POMDP actions include $m_1, m_2, \ldots$, and the POMDP observations are, for instance, the ASR output utterances, i.e., $\tilde{u}_1, \tilde{u}_2, \ldots$, or the keywords extracted from the ASR output utterances. At any case, the observations provide only partial information about the POMDP states, i.e., the user intents.

The transition model is a probability model representing uncertainty in the effect of system's actions, and it needs to be learned ideally from dialogs. For example, the transition model can encode the probability that the users change their intents between the dialog turns after receiving the system's actions. The observation model is a probability model for uncertainty in the domain. For instance, the probability that a particular keyword is an indicator for a particular state, say the probability that the keyword *restaurant* is an indicator for the state *food* places.

The POMDP reward function encodes the immediate reward for the system's executing an action in a state. The reward function, which can also be considered as a cost function, is the most succinct element that encodes the performance of the system. For example, in the dialog POMDPs the reward function is usually *defined* as: (a) a small negative number (for instance $-1$) for each action of the system at any dialog turn, (b) a large positive reward (for instance $+10$) if the dialog ends successfully, and (c) a large negative reward (for instance $-100$) otherwise.

Given a POMDP model, we can apply dynamic programming techniques to solve the POMDP, i.e., to find the (near) *optimal* policy (Cassandra et al. 1995). The optimal policy is the policy that optimizes the reward function for any dialog state sequence. The POMDP's (near) optimal policy, shortly called the POMDP policy, represents the dialog manager's *strategy* for any dialog situation. That is, the dialog manager performs the best action at any dialog state based on the optimized policy. Thus, we use the words policy and strategy interchangeably.

Estimating the model components of dialog POMDPs is a significant issue; as the POMDP model has direct impact on the POMDP policy and consequently on the applicability of the POMDP in the domain of interest. In this context, the SDS researchers in both academia and industry have addressed several practical challenges of applying POMDPs to SDS (Roy et al. 2000; Williams 2006; Paek and Pieraccini 2008). In particular, learning the SDS dynamics ideally from the available unannotated and noisy dialogs is a challenge for us.

In many real applications including SDSs, it is usual to have large amount of unannotated data, such as web-based spoken query retrieval (Ko and Seo 2004). Manually annotating the data is an expensive task, thus learning from unannotated data is an interesting challenge which is tackled using unsupervised learning methods. Therefore, we are interested in learning the POMDP model components based on the available unannotated data.

Furthermore, POMDPs, unlike MDPs, have scalability issues. That is, finding the (near) optimal policy of the POMDP highly depends on the number of states, actions and observations. In particular, the number of observations can exponentially increase the number of conditional plans (Kaelbling et al. 1998). For example, in most non-trivial dialog domains, the POMDP model can include hundreds or thousands of observations such as words or user utterances. In the example given in Table 1.1, $\tilde{u}_1$, $\tilde{u}_2$, $\tilde{u}_3$, and $\tilde{u}_4$, together with many other possible utterances, can be considered as observations. Finding the optimal policy of such a POMDP is basically intractable.

Finally, as mentioned above, the reward function of a POMDP highly affects the optimized policy. The reward function is perhaps the most hand-crafted aspect of the optimization frameworks such as POMDPs (Paek and Pieraccini 2008). Using inverse reinforcement learning (IRL) (Ng and Russell 2000), a reward function can be determined from behavioral observation. Fortunately, learning the reward function using IRL methods has been already proposed for the general POMDP framework (Choi and Kim 2011), paving the path for investigating its use for dialog POMDPs.

**Algorithm 1:** The descriptive algorithm to learn the dialog POMDP model components using unannotated dialogs

**Input**: The unannotated dialog set of interest
**Output**: The learned dialog POMDP model components that can be used in a POMDP solver to find the (near) optimal policy

1. *Learn the dialog intents from unannotated dialogs using an unsupervised topic modeling approach, and make use of them as the dialog POMDP states;*
2. *Extract actions directly from dialogs and learn a maximum likelihood transition model using the learned states;*
3. *Reduce observations significantly and learn the observation model;*
4. *Learn the reward function based on the IRL technique and using the learned POMDP model components;*

## 1.1 An End-to-End Approach

In this book, we propose an end-to-end approach for learning the model components of dialog POMDPs from unannotated and noisy dialogs of intent-based dialog domains (Chinaei 2013). The big picture of this work is presented in the descriptive Algorithm 1. The input to the algorithm is any unannotated dialog set. In this book, we use SACTI-1 dialog data (Williams and Young 2005) and SmartWheeler dialogs (Pineau et al. 2011).

In step 1, we address learning the dialog intents from unannotated dialogs using an unsupervised topic modeling approach, and make use of them as the dialog POMDP states. In step 2, we directly extract the actions from the dialog set and learn a maximum likelihood transition model using the learned states. In step 3, we reduce observations significantly and learn the observation model. Specifically, we propose two observation models: the keyword model and the intent model.

Building on the learned dialog POMDP model components, we propose two IRL algorithms for learning the reward function of dialog POMDPs from dialogs, in step 4. The learned reward function makes the dialog POMDP model complete. The learned dialog POMDP model can be used in an available model-based POMDP solver to find the optimal policy.

In this book, we present several illustrative examples. We use SACTI-1 dialogs to run the proposed methods and show the results throughout the book. In the end, we apply the proposed methods on healthcare dialog management in order to learn a dialog POMDP from dialogs collected by an intelligent wheelchair, called SmartWheeler.

To the best of our knowledge, this is the first work that proposes and implements an end-to-end learning approach for dialog POMDP model components. That is, starting from scratch, it learns the state, the transition model, the observations and the observation model, and finally the reward function. These altogether form a significant set of contributions that can potentially inspire substantial further work.

The rest of the book is organized as follows. We describe the necessary background knowledge in Chaps. 2 and 3. In particular, in Chap. 2, we introduce topic modeling that is used for intent learning later in the book. In Chap. 3, we introduce the MDP and POMDP frameworks. In Chap. 4 we go through steps 1–3 in the descriptive Algorithm 1. That is, we describe the methods for learning more basic model components of dialog POMDPs: the states and transition model, the observations and observation model. Then in Chap. 5, we review IRL in the MDP framework followed by our proposed POMDP-IRL algorithms for learning the reward function in dialog POMDPs. In Chap. 6, we apply the whole methods on SmartWheeler, to learn a dialog POMDP from SmartWheeler dialogs. Finally, we conclude and address the future work in Chap. 7.

# Chapter 2
# A Few Words on Topic Modeling

Topic modeling techniques are used to discover the topics for (unlabeled) texts. As such, they are considered as unsupervised learning techniques which try to learn the patterns inside the text by considering words as observations. In this context, latent Dirichlet allocation (LDA) is a Bayesian topic modeling approach which has useful properties particularly for practical applications (Blei et al. 2003). In this section, we go through LDA by first reviewing the Dirichlet distribution, which is the basic distribution used in LDA.

## 2.1 Dirichlet Distribution

Dirichlet distribution is the *conjugate* prior for multinomial distribution likelihood (Kotz et al. 2000; Balakrishnan and Nevzorov 2003; Fox 2009). Specifically, the conjugate prior of a distribution has the property that after updating the prior, the posterior also has the same functional form as the prior (Hazewinkel 2002; Robert and Casella 2005). It has been shown that conjugate priors are found only inside the exponential families (Brown 1986).

### *2.1.1 Exponential Distributions*

The density function of exponential distributions has a factor called *sufficient statistic*. The sufficient statistic is the sufficient function of the sample data (as reflected by its name) such that no other statistic that can be calculated from the sample



H. Chinaei, B. Chaib-draa, *Building Dialogue POMDPs from Expert Dialogues*,
SpringerBriefs in Speech technology, DOI 10.1007/978-3-319-26200-0_2

data provides any additional information than the sufficient statistic (Fisher 1922; Hazewinkel 2002). For instance, the maximum likelihood estimator in exponential families depends on the sufficient statistic but not all of observations.

The exponential families have the property that the dimension of sufficient statistic is bounded even if the size of observations goes to infinity, except a few member of exponential families such as uniform distribution. Moreover, the important property of exponential families is inside the theorems independently proved by Pitman (1936), Koopman (1936), and Darmois (1935) approximately at the same time. This property leads to efficient parameter estimation methods in exponential families. Examples of exponential families are the normal, Gamma, Poison, multinomial, and Dirichlet distributions. In particular, the Dirichlet distribution is the *conjugate* prior for the multinomial distribution likelihood.

### 2.1.2 Multinomial Distribution

For the multinomial distribution, consider the trial of $n$ events with observations $\boldsymbol{y} = (y_1, \ldots, y_n)$ and the parameters $\boldsymbol{\pi} = (\pi_1, \ldots, \pi_k)$ where the observation of each event can take $K$ possible values. For instance, in events of rolling a fair die $n$ times, each observation $y_i$ can take $K = 6$ values with equal probabilities $\left(\pi_1 = \frac{1}{6}, \ldots, \pi_k = \frac{1}{6}\right)$. Under such condition, this experiment is governed by a multinomial distribution. Formally, for the probability of having an observation $\boldsymbol{y} = (y_1, \ldots, y_n)$ given the parameters $\boldsymbol{\pi} = (\pi_1, \ldots, \pi_k)$ we have:

$$p(\boldsymbol{y}|\boldsymbol{\pi}) = \frac{n!}{\prod_{i=1}^{K} n_i!} \prod_{i=1}^{K} \pi_i^{n_i},$$

where

$$n_i = \sum_{j=1}^{n} \delta(y_j, i)$$

in which $\delta(x, y)$ is the Kronecker delta function; $\delta(x, y) = 1$ if $x = y$, and zero otherwise. Moreover, it can be shown that in multinomial distribution, the expectation of number of times that the value $i$ is observed over $n$ trials is:

$$E(Y_i) = n\pi_i$$

and its variance is:

$$\text{Var}(Y_i) = n\pi_i(1 - \pi_i).$$

### 2.1.3 Dirichlet Distribution

For the conjugate prior of the likelihood of multinomial distribution, i.e., $p(\boldsymbol{\pi}|\mathbf{y})$, assume that the prior $p(\boldsymbol{\pi} = (\pi_1, \ldots, \pi_k))$ is drawn from Dirichlet distribution with the hyper parameters $\boldsymbol{\alpha} = (\alpha_1, \ldots, \alpha_k)$ then the posterior $p(\boldsymbol{\pi}|\mathbf{y})$ is also drawn from Dirichlet distribution with the hyper parameters $(\alpha_1 + n_1, \ldots, \alpha_k + n_k)$. Recall that $n_i$ is the number of times the value $i$ has been observed in the last trial, where $1 \leq i \leq K$.

This is the useful property of Dirichlet distribution which says that for updating the prior to get the posterior it suffices only to update the hyper parameters. That is, having a Dirichlet prior with the hyper parameters $\boldsymbol{\alpha} = (\alpha_1, \ldots, \alpha_k)$, after observing observations $(n_1, \ldots, n_k)$ the posterior hyper parameters become $(\alpha_1 + n_1, \ldots, \alpha_k + n_k)$. This property is discussed in the illustrative example further in this section.

Then, Dirichlet distribution for the parameter $\boldsymbol{\pi}$ with hyper parameter $\boldsymbol{\alpha}$ would be:

$$p(\boldsymbol{\pi}|\boldsymbol{\alpha}) = \frac{\Gamma(\sum_i(\alpha_i))}{\prod_i(\Gamma(\alpha_i))} \prod_{i=1}^{K} \pi_i^{\alpha_i - 1},$$

where $\Gamma(x)$ is the standard Gamma function. Note that Gamma function is an extension of factorial function. That is, for positive numbers Gamma function is the factorial function, i.e., $\Gamma(n) = n!$. Moreover, it can be shown that the expectation of Dirichlet prior $\pi$ is:

$$E(\pi_i) = \frac{\alpha_i}{s} \tag{2.1}$$

and its variance is:

$$\mathrm{Var}(\pi_i) = \frac{E(\pi_i)(1 - E(\pi_i))}{s + 1},$$

where $s = \alpha_1 + \cdots + \alpha_k$ and is called the *concentration parameter*. The concentration parameter controls how concentrated the distribution is around its expected value (Sudderth 2006). The higher $s$ is, the lower is the variance of the parameters. Moreover, given the concentration parameter $s$, the higher the hyper $\alpha_i$ is, the higher the expected value of $\pi_i$ is. Therefore, the Dirichlet hyper parameters $\boldsymbol{\alpha} = (\alpha_1, \ldots, \alpha_i)$ operate as a confidence measure.

Figure 2.1 plots 3 Dirichlet distributions with 3 values for $s$ in three unit simplex (with 3 vertices). Note that $p(\boldsymbol{\pi})$ is a point in each simplex and $0 \leq \pi_i$, and $\sum_i^K \pi_i = 1$. Figure 2.1 shows that the higher the $s$ is, the more concentration is around its expected value. In addition, the simplex in the middle has a high $s$ whereas the one in the right has a lower $s$.

Neapolitan (2004) proved the useful property for the posterior of Dirichlet distribution. Suppose we are about to repeatedly perform an experiment with $k$

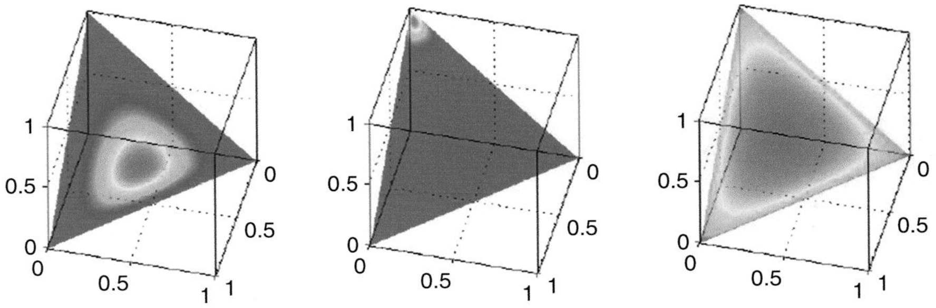


**Fig. 2.1** The Dirichlet distribution for different values of the concentration parameter, taken from Huang (2005)

outcomes $x_1, x_2, \ldots, x_k$. We assume exchangeable observations and present our prior belief concerning the probability of heads using a Dirichlet distribution with the parameters $\boldsymbol{\alpha} = (\alpha_1, \ldots, \alpha_k)$. Then, our prior probabilities become:

$$p(x_1) = \frac{\alpha_1}{m} \; \ldots \; p(x_k) = \frac{\alpha_k}{m},$$

where $m = \alpha_1 + \cdots + \alpha_k$.

After observing $x_1, \ldots, x_k$ occurs, respectively, $n_1, \ldots, n_k$ times in $n$ trials where $n = n_1 + \cdots + n_k$. Then, our posterior probabilities become as follows:

$$\begin{aligned} p(x_1|n_1, \ldots, n_k) &= \frac{\alpha_1 + n_1}{s = m + n} \\ &\ldots \\ p(x_k|n_1, \ldots, n_k) &= \frac{\alpha_k + n_k}{s = m + n}. \end{aligned} \tag{2.2}$$

### 2.1.4 Example on the Dirichlet Distribution

Here, we present an illustrative example for the Dirichlet distribution, taken from Neapolitan (2009). Suppose we have an asymmetrical, six-sided die, and we have little idea of the probability of each side coming up. However, it seems that all sides are equally likely. So, we assign equal initial confidence about observing each number 1–6 appear by the die on the Dirichlet hyper parameters $\boldsymbol{\alpha} = (\alpha_1, \ldots, \alpha_k)$ as follows:

$$\alpha_1 = \alpha_2, \ldots, \alpha_6 = 3.$$

**Table 2.1** Dirichlet distribution example: the results of throwing a die 100 times

| Outcome ($x_i$) | Number of occurrences ($n_i$) |
|---|---|
| 1 | 10 |
| 2 | 15 |
| 3 | 5 |
| 4 | 30 |
| 5 | 13 |
| 6 | 27 |
| $n$ | 100 |

**Table 2.2** Dirichlet distribution example: the updated posterior probabilities

| |
|---|
| $p(1\|10, 15, 5, 30, 13, 27) = \frac{\alpha_1+n_1}{s} = \frac{3+10}{18+100} = 0.110$ |
| $p(2\|10, 15, 5, 30, 13, 27) = \frac{\alpha_1+n_1}{s} = \frac{3+15}{18+100} = 0.153$ |
| $p(3\|10, 15, 5, 30, 13, 27) = \frac{\alpha_1+n_1}{s} = \frac{3+5}{18+100} = 0.067$ |
| $p(4\|10, 15, 5, 30, 13, 27) = \frac{\alpha_1+n_1}{s} = \frac{3+30}{18+100} = 0.280$ |
| $p(5\|10, 15, 5, 30, 13, 27) = \frac{\alpha_1+n_1}{s} = \frac{3+13}{18+100} = 0.136$ |
| $p(6\|10, 15, 5, 30, 13, 27) = \frac{\alpha_1+n_1}{s} = \frac{3+27}{18+100} = 0.254$ |

**Table 2.3** Dirichlet distribution example: the updated hyper parameters

| |
|---|
| $\alpha_1 = \alpha_1 + n_1 = 3 + 10 = 13$ |
| $\alpha_2 = \alpha_2 + n_2 = 3 + 15 = 18$ |
| $\alpha_3 = \alpha_3 + n_3 = 3 + 5 = 8$ |
| $\alpha_4 = \alpha_4 + n_4 = 3 + 30 = 33$ |
| $\alpha_5 = \alpha_5 + n_2 = 3 + 13 = 16$ |
| $\alpha_6 = \alpha_6 + n_2 = 3 + 27 = 30$ |

Then, we have $s = 3 \times 6 = 18$, and the prior probabilities are as follows:

$$p(1) = p(2) = \ldots = p(6) = \frac{\alpha_i}{s} = \frac{3}{18} = 0.16667.$$

Next, suppose that we throw the die 100 times, with the following results shown in Table 2.1.

Using Eq. (2.2), the posterior probabilities can be updated as shown in Table 2.2.

Note in the example that the new value for the concentration parameter becomes $s = m+n$, where $m = 18$ $(\alpha_1 + \cdots + \alpha_k)$, and $n = 100$ (the number of observations). Moreover, the new values of hyper parameters become as shown in Table 2.3.

Using Eq. (2.1), $E(\pi_i) = \alpha_i/s$, the expected value of the parameters can be calculated as shown in Table 2.4.

Comparing the values in Table 2.4 to the ones in Table 2.2, we can see another important property of the Dirichlet distribution. That is, the number of observations directly reveals the confidence on the expected value of parameters.

In this section, we observed the Dirichlet distribution's useful properties:

1. The Dirichlet distribution is the conjugate prior for likelihood of multinomial distribution,

**Table 2.4** Dirichlet distribution example: the expected value of hyper parameters

| |
|---|
| $E(\pi_1) = \alpha_1/s = 13/118 = 0.110$ |
| $E(\pi_2) = \alpha_2/s = 18/118 = 0.153$ |
| $E(\pi_3) = \alpha_3/s = 8/118 = 0.280$ |
| $E(\pi_4) = \alpha_4/s = 33/118 = 0.067$ |
| $E(\pi_5) = \alpha_5/s = 16/118 = 0.136$ |
| $E(\pi_6) = \alpha_6/s = 30/118 = 0.254$ |

**Table 2.5** The LDA example: the given text

| | |
|---|---|
| 1: | I eat orange and apple since those are juicy. |
| 2: | The weather is so windy today. |
| 3: | The hurricane Catherine passed with no major damage. |
| 4: | Watermelons here are sweat because of the hot weather. |
| 5: | Tropical storms usually end by November. |

2. For updating the posterior of multinomial distribution with Dirichlet prior, we need only to update the Dirichlet prior by adding the observation counts to the Dirichlet hyper prior, and
3. The number of observations directly reveals the confidence on the expected value of the parameters.

Because of these important properties, the Dirichlet distribution is applied largely in different applications. In particular, LDA assumes that the learned parameters follow the Dirichlet distribution. The following section describes the LDA method.

## 2.2 Latent Dirichlet Allocation

LDA is a latent Bayesian topic model which is used for discovering the hidden topics of documents (Blei et al. 2003). In this model, a document can be represented as a mixture of the hidden topics, where each hidden topic is represented by a distribution over words occurred in the document. Suppose we have the sentences shown in Table 2.5.

Then, the LDA method automatically discovers the topics that the given text contain. Specifically, given two asked topics, LDA can learn the two topics and the topic assignments to the given text. The learned topics are represented using the words and their probabilities of occurring for each topic as presented in Table 2.6. The topic representation for topic A illustrates that this topic is about *fruits*. And, the topic representation for topic B illustrates that this topic B is about the *weather*. Then, the topic assignment for each sentence can be calculated as presented in Table 2.7.

Formally, given a document in the form of $\boldsymbol{d} = (w_1, \ldots, w_M)$ in a document corpus (set), $\boldsymbol{D}$, and given $N$ asked topics, the LDA model learns two parameters:

1. The parameter $\boldsymbol{\theta}$ which is generated from the Dirichlet prior $\boldsymbol{\alpha}$.
2. The parameter $\boldsymbol{\beta}$ which is generated from Dirichlet prior $\boldsymbol{\eta}$.

**Table 2.6** The LDA example: the learned topics

| Topic | A (%) | Topic | B (%) |
|---|---|---|---|
| Orange | 20 | Weather | 30 |
| Apple | 20 | Windy | 10 |
| Juicy | 5 | Hot | 10 |
| Sweat | 1 | Storm | 9 |
| ... | ... | ... | ... |

**Table 2.7** The LDA example: the topic assignments to the text

| | | | | |
|---|---|---|---|---|
| Sentence 1 | Topic A | 100 % | Topic B | 0 % |
| Sentence 2 | Topic A | 0 % | Topic B | 100 % |
| Sentence 3 | Topic A | 0 % | Topic B | 100 % |
| Sentence 4 | Topic A | 65 % | Topic B | 35 % |
| Sentence 5 | Topic A | 0 % | Topic B | 100 % |

The first parameter, $\boldsymbol{\theta}$, is a vector of size $N$ for distribution of hidden topics, $z$. The second one, $\boldsymbol{\beta}$, is a matrix of size $M \times N$ in which the column $j$ stores the probability of each word given the topic $z_j$.

Figure 2.2 shows the LDA model in the plate notation in which the boxes are plates, that represents replicates. The shaded nodes are the observation nodes, i.e., the words $w$. The unshaded nodes $z$ represent hidden topics. Then, the generative model of LDA performs as follows:

1. For each document, $\boldsymbol{d}$, a parameter, $\boldsymbol{\theta}$, is drawn for the distribution of hidden topics based on multinomial distribution with the Dirichlet parameters $\boldsymbol{\alpha}$ (cf. Dirichlet distribution in Sect. 2.1).
2. For each document set $\boldsymbol{D}$, a parameter, $\boldsymbol{\beta}$, is learned for the distribution of words given topics. Given each topic $z$, the vector $\boldsymbol{\beta}_z$ is drawn based on multinomial distribution with the Dirichlet parameters $\boldsymbol{\eta}$.
3. Generate the $j$th word in the document $i$, $w_{i,j}$, as:
   (a) Draw a topic $z_{i,j}$ based on the multinomial distribution with the parameter $\theta_i$.
   (b) Draw a word based on the multinomial distribution with the parameter $\phi_{z_{i,j}}$.

## *Comparison to Earlier Models*

Blei et al. (2003) compared the LDA model to the related earlier models such as unigrams and mixture of unigrams (Bishop 2006; Manning and Schütze 1999), as well as probabilistic latent semantic analysis (PLSA) (Hofmann 1999). These three models are represented in Fig. 2.3.

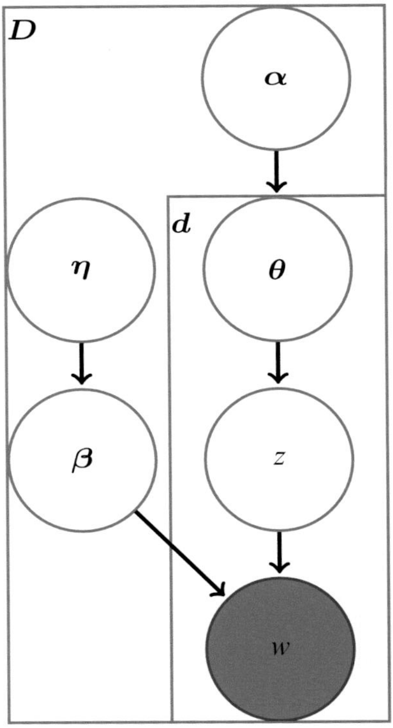


**Fig. 2.2** Latent Dirichlet allocation

Figure 2.3a shows the unigram model. In unigrams, a document $\boldsymbol{d} = (w_1, \ldots, w_n)$ is a mixture of words. So, the probability of having a document $\boldsymbol{d}$ is calculated as:

$$p(\boldsymbol{d}) = \prod_{w_i} p(w_i)$$

Then, in the mixture of unigrams in Fig. 2.3b, a word $w$ is drawn from a topic $z$ this time. Under this model, a document $\boldsymbol{d}$ is generated by:

1. Draw a hidden topic $z$.
2. Draw each word $w$ based on the hidden topic $z$.

As such, in mixture of unigrams the probability of having the document $\boldsymbol{d}$ is calculated as:

$$p(\boldsymbol{d}) = \sum_{z} p(z) \prod_{w_i} p(w_i|z).$$

Notice that mixture of unigrams assumes that each document $\boldsymbol{d}$ includes only one hidden topic. This assumption is removed in PLSA model shown in Fig. 2.3c. In PLSA, a distribution $\boldsymbol{\theta}$ is sampled and attached to each *observed* document

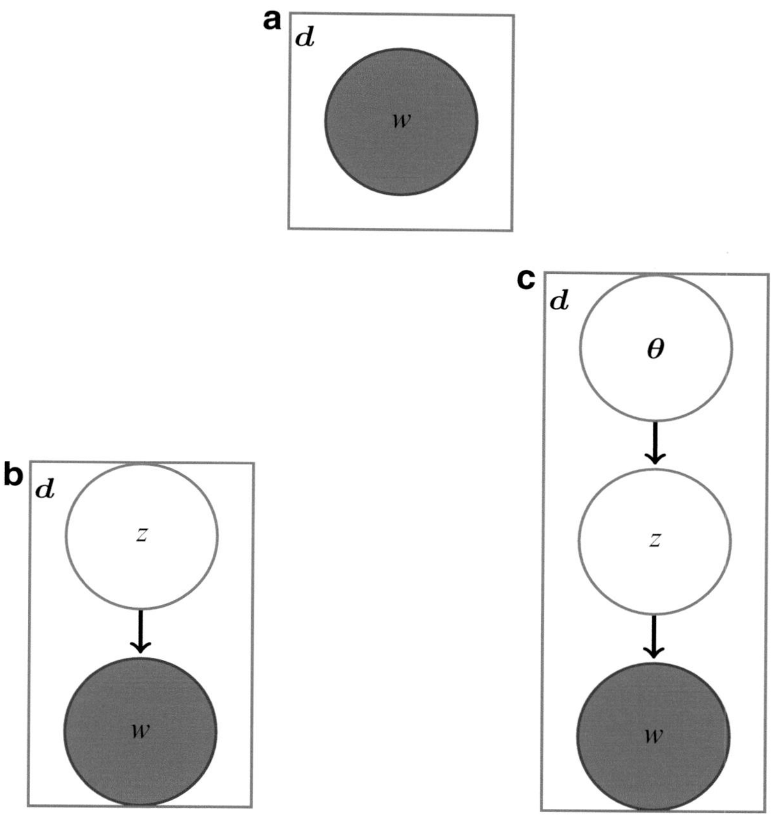


**Fig. 2.3** (**a**) Unigram model. (**b**) Mixture of unigrams. (**c**) Probabilistic latent semantic analysis (PLSA)

for the distribution of hidden topics. Then, the probability of having a document $\boldsymbol{d} = (w_1, \ldots, w_n)$ is calculated as:

$$p(\boldsymbol{d}) = \sum_z p(z|\boldsymbol{\theta}) \prod_{w_i} p(w_i|z),$$

where $\boldsymbol{\theta}$ is the distribution of hidden topics.

Note also that LDA is similar to PLSA in that both LDA and PLSA learn a parameter $\boldsymbol{\theta}$ for the distribution of hidden topics of each document. Then, the probability of having a document $\boldsymbol{d} = (w_1, \ldots, w_n)$ is calculated using:

$$p(\boldsymbol{d}) = \sum_z p(z|\boldsymbol{\theta}) \prod_{w_i} p(w_i|z),$$

where $\boldsymbol{\theta}$ is the distribution of hidden topics.

In contrast to PLSA, in LDA first a parameter $\boldsymbol{\alpha}$ is generated which is used as the Dirichlet prior for the multinomial distribution $\boldsymbol{\theta}$ of topics. In fact, Dirichlet prior can be used as a natural way to assign more probability to the random variables on which we have more confidence. Moreover, use of Dirichlet prior leads to interesting advantages of LDA over PLSA. First, as opposed to PLSA, LDA does not require to visit a document $\boldsymbol{d}$ to sample a parameter $\boldsymbol{\theta}$. But in LDA, the parameter $\boldsymbol{\theta}$ is generated using the Dirichlet parameter $\boldsymbol{\alpha}$. As such, LDA is a well-defined generative model of documents which is able to assign probabilities to a previously *unseen* document of the corpus. Moreover, LDA is not dependent on the size of corpus and does not overfit as opposed to PLSA (Blei et al. 2003).

So, LDA is a topic modeling approach that considers mixture of hidden topics for documents, where documents are seen as bag of words. However, it does not consider the Markovian property among sentences. Later in this book, we introduce a variation of LDA that adds the Markovian property to LDA, for the topic transition from one sentence to the following one. In this context, hidden Markov models (HMMs) are used for modeling Markovian property particularly in texts. In the following section, we briefly review HMMs.

## 2.3 Hidden Markov Models

In Markovian domains the current environment's state depends on the state in the previous time step, similar to finite state machines. In fact, Markov models are generalized models of finite state machines in which the transitions are not deterministic. That is, in Markov models the current environment state depends on the previous state and the probability of landing to the current state, known as the transition probability (Manning and Schütze 1999).

In hidden Markov models (HMMs) (Rabiner 1990), as opposed to Markov models, states are not fully observable, but there is the idea of observations which give the current state of the model with only some probability. So, in HMMs there is an observation model besides the transition model. Similar to the Markov models, in HMMs the transition model is used for estimating the current state of the model with some probability, given the previous state. As such, we can state that an HMM with a deterministic observation model is equivalent to a Markov model, and that a Markov model with a deterministic transition model is equivalent to a finite state machine.

Figure 2.4 shows an HMM where hidden states $s_1, \ldots, s_n$ are inside circles and observations $o_1, \ldots, o_n$ are noted inside the shaded circles. The Markovian property in HMMs states that at each time step the state of the HMM depends on its previous state $p(s_t|s_{t-1})$, and the current observation depends on the current state $p(o_t|s_t)$.

Formally, an HMM is defined as a tuple $(S, O, A, B, \Pi)$:

- $S = s_1, \ldots, s_N$ is a set of $N$ states,
- The transition probability matrix $A$

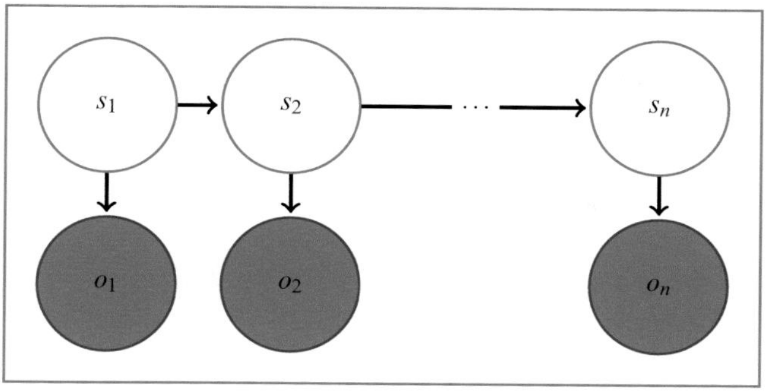


**Fig. 2.4** The hidden Markov model, the *shaded nodes* are observations ($o_i$) used to capture hidden states ($s_i$)

$$A = \begin{pmatrix} a_{11}, \ldots, a_{1n} \\ \ldots \\ a_{n1}, \ldots, a_{nn} \end{pmatrix}.$$

Each $a_{ij}$ represents the probability of moving from state $i$ to state $j$, s.t. $\sum_{j=1}^{n} a_{ij} = 1$,

- $O = o_1 o_2 \ldots o_T$, is sequence of $T$ observations, each one drawn from a vocabulary $V = v_1, v_2, \ldots, v_V$,
- $B = b_i(o_t)$, is a sequence of observation likelihoods, also called emission probabilities, each expressing the probability of an observation $o_t$ being generated from a state $i$,
- $\Pi$ is the initial probability model which shows the probability that the model starts with each state in $S$.

Then, there are three fundamental questions that we want to answer in HMMs (Jurafsky and Martin 2009; Manning and Schütze 1999):

1. The first problem is to compute the *likelihood* of a particular observation sequence. Formally, we want to find out:

   Given an HMM, $\lambda = (A, B)$ and an observation sequence $O$, determine the likelihood $\Pr(O|\lambda)$.
2. Learning the most likely state sequence given a sequence of observations and the model. This problem is called *decoding*. This is interesting, for instance, in part-of-speech tagging where given a set of words as observations we would like to infer about the most probable tags of the words (Church 1988). Formally, we want to find out:

   Given as input and HMM $\lambda = (A, B)$, and a sequence of observations $O = o_1, o_2, \ldots, o_T$, find the most probable sequence of states, i.e., $(s_1, \ldots, s_T)$.

   That is, we want to find out *the state sequence that best explains the observations.*

3. HTMM training, i.e., learning the HMM parameters. Given a sequence of observations what the most probable model parameters are:

$$\underset{\lambda}{\operatorname{argmax}}\, p(o_1, \ldots, o_n|\lambda). \tag{2.3}$$

This problem is called *parameter estimation.*

Note that there is no analytical solution for the maximization of parameter estimation in Eq. (2.3). This problem is tackled with a well-known algorithm named as *Baum–Welch* or *Forward–Backward* algorithm (Welch 2003), which is an *Expectation Maximization* (EM) algorithm.

In fact, EM is a class of algorithms for learning unknown parameters of a model. The basic idea is to pretend that the parameters of the model are known and then to infer the probability that each observation belongs to each model (Russell and Norvig 2010). Then, the model refits to the observations. Specifically, each model is fitted to all observations, and each observation is weighted by the probability that it belongs to that model. This process iterates until convergence.

EM algorithms start with a random parameter, and calculate the probability of observations. Then, they observe in the calculations to find which state transitions and observation probabilities have been used most, and increase the probability of those. This process leads to an updated parameter which gives higher probability to the observations. Then, the following two steps are iterated until convergence: calculating the probabilities of observations given a parameter (expectation) and updating the parameter (maximization).

Formally, an EM algorithm works as follows. Assuming the set of parameter $\Theta$, hidden variables $Z$ and observations $X$. First, the function $Q$ is defined as (Dempster et al. 1977):

$$Q(\Theta|\Theta^t) = E[\log p(X, Z|\Theta)|X, \Theta^t]. \tag{2.4}$$

Then, in the expectation and maximization steps the following calculations are performed:

1. Expectation: $Q(\Theta|\Theta^t)$ is computed.
2. Maximization: $\Theta^{t+1} = \operatorname{argmax}_\Theta Q(\Theta|\Theta^t)$
   That is, the parameter $\Theta^{t+1}$ is set to the $\Theta$ that maximizes $Q(\Theta|\Theta^t)$.

For instance, in Baum–Welch algorithm the expectation and maximization steps are as follows:

1. In the expectation the following two calculations are done:
   - Calculating the expected number of times that observation $o$ has been observed from state $s$ for all states and observations, given the current parameter of the model.
   - Calculating the expected number of times that state transitions from state $s_i$ to state $s_j$ is done, given the current parameters of the model.

2. In the maximization step the parameters $A$, $B$, and $\Pi$ are set to the parameters which maximize the expectations above.

More specifically, the Expectation and Maximization step for HMM parameter learning can be derived as described in Jurafsky and Martin (2009):

1. Expectation:

$$\gamma_t(j) = \frac{\alpha_t(j)\beta_t(j)}{\Pr(O|\gamma)} \quad \forall\ t \text{ and } j$$

$$\xi_t(i,j) = \frac{\alpha_t(i)a_{ij}b_j(o_{t+1})\beta_{t+1}(j)}{\alpha_T(N)} \quad \forall\ t,\ i, \text{ and } j,$$

where $\alpha_t$ is known as *forward path probability*:

$$\alpha_t(j) = \Pr(o_1, o_2, \ldots, o_t, s_t = j|\lambda)$$

and $\beta_t(j)$ is known as *backward path probability*:

$$\beta_t(i) = \Pr(o_{t+1}, o_{t+2}, \ldots, o_T|s_t = i, \lambda)$$

2. Maximization:

$$\hat{a}_{i,j} = \frac{\sum_{t=1}^{T-1} \xi(i,j)}{\sum_{t=1}^{T-1} \sum_{j=1}^{N} \xi(i,j)}$$

$$\hat{b}_j(v_k) = \frac{\sum_{t=1 s.t. O_t = v_k}^{T} \gamma_t(j)}{\sum_{t=1}^{T} \gamma_t(j)}.$$

In this section, we introduced the basic methods used in topic modeling. In particular, we studied the LDA method and HMMs, the background for hidden topic Markov model (HTMM). The HTMM approach adds Markovian property to the LDA method, and is introduced in Chap. 4. In the following chapter, we introduce the sequential decision making domain and its application on spoken dialogue systems.

# Chapter 3
# Sequential Decision Making in Spoken Dialog Management

This chapter includes two major sections. In Sect. 3.1, we introduce sequential decision making and study the supporting mathematical framework for it. We describe the Markov decision process (MDP) and the partially observable MDP (POMDP) frameworks, and present the well-known algorithms for solving them. In Sect. 3.2, we introduce spoken dialog systems (SDSs). Then, we study the related work of sequential decision making in spoken dialog management. In particular, we study the related research on application of the POMDP framework for spoken dialog management. Finally, we review the user modeling techniques that have been used for dialog POMDPs.

## 3.1 Sequential Decision Making

In sequential decision making, an agent needs to take sequential actions, during the interaction with an environment. The agent's interaction with the environment can be in a stochastic and/or uncertain situation. That is, the effect of the actions is not completely known (in stochastic domains) and observations from the environment provide incomplete or error-prone information (in uncertain domains). As such, sequential decision making under such condition is a challenging problem.

Figure 3.1 shows the cycle of interaction between an agent and its environment. The agent performs an *action* and receives an *observation* in return. The observation can be used by the agent, for instance to update its *state* and *reward*. The reward works as a reinforcement from the environment that shows how well the agent performed. In sequential decision making, the agent is required to make decision for sequence of states rather than making a one-shot decision. Then, the sequential



H. Chinaei, B. Chaib-draa, *Building Dialogue POMDPs from Expert Dialogues*, SpringerBriefs in Speech technology, DOI 10.1007/978-3-319-26200-0_3

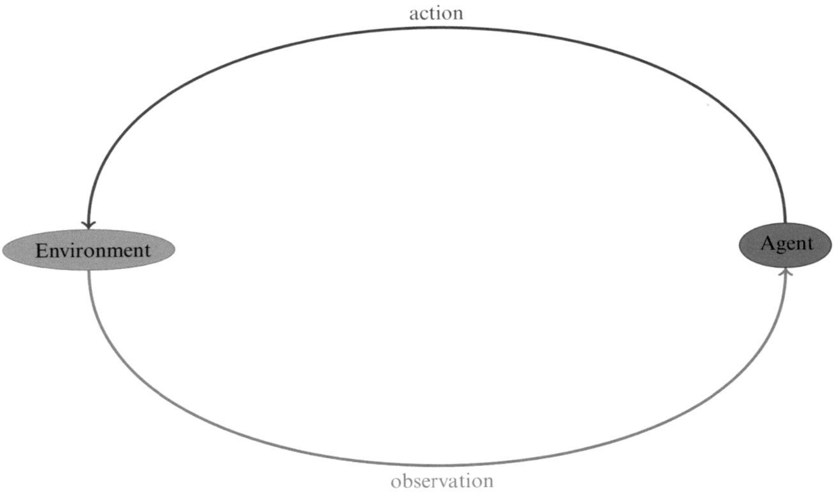


**Fig. 3.1** The cycle of interaction between an agent and the environment

decision making is performed with the objective of *maximizing* the long-term rewards. The sequence of actions is called a *policy*, and the major question in sequential decision making is how to find a near optimal policy.

In stochastic domains where the decision making is sequential, the suitable formal framework to find the near optimal policy is the MDP. However, the MDP framework considers the environment as fully observable and this does not conform to real applications which are partially observable such as SDSs. In this context, the POMDP framework can deal this constraint of uncertainty. The MDP/POMDP frameworks are composed of model components which can be used, for instance, for representing the available stochasticity and uncertainty.

If the MDP/POMDP model components are *not* known in advance, then *reinforcement learning (RL)* is used to learn the near optimal policy. In fact, RL is a series of techniques in which the agent learns the near optimal policy in the environment based on the agent's own experience (Sutton and Barto 1998). The better the agent acts, the more rewards it achieves. Then, the agent aims to maximize its expected rewards over time. Since in RL the model components are usually unknown, RL is called model-free RL; particularly in spoken dialog community (Rieser and Lemon 2011).

On the other hand, if the model components of the underlying MDP/POMDP framework are known in advance, then we can *solve* MDPs/POMDPs, which is a search through the state space for an optimal policy or path to goal using the available planning algorithms (Bellman 1957a). This method is also called model-based RL, particularly in the spoken dialog community (Rieser and Lemon 2011).

In this book, we are interested in learning the environment dynamics of a dialog manager in advance and make use of them in the POMDP model components. We then refer to such dialog manager as *dialog POMDP*. Once the dialog POMDP model components are learned, we can solve the POMDP for the optimal policy using the available planning algorithms. In the following section, we introduce the MDP and POMDP background.

### 3.1.1 Markov Decision Processes

A MDP is a mathematical framework for decision making under uncertainty (Bellman 1957b). An MDP is defined as $(S, A, T, R, \gamma, s_0)$, where

- $S$ is the set of discrete states,
- $A$ is the set of discrete actions,
- $T$ is the transition model which consists of the probabilities of state transitions:

$$T(s, a, s') = \Pr(s_{t+1} = s' | a_t = a, s_t = s),$$

where $s$ is the current state and $s'$ is the next state,
- $R(s, a)$ is the reward of taking action $a$ in the state $s$,
- $\gamma$ is the discount factor, a real number between 0 and 1,
- and $s_0$ is an initial state.

Then, a *policy* is the selection of an action $a$ in a state $s$. That is, the policy $\pi$ maps each state $s$ to an action $a$, i.e., $a = \pi(s)$. In an MDP, the objective is to find an optimal policy $\pi^*$, that maximizes the value function, i.e., the expected discount of future rewards starting from state $s_0$:

$$V^{\pi}(s) = E_{s_t \sim T}\Big[\gamma^0 R(s_0, \pi(s_0)) + \gamma^1 R(s_1, \pi(s_1)) + \cdots | \pi, s_0 = s\Big]$$

$$V^{\pi}(s) = E_{s_t \sim T}\Big[\sum_{t=0}^{\infty} \gamma^t R(s_t, \pi(s_t)) | \pi, s_0 = s\Big].$$

The value function of a policy can also be recursively defined as:

$$\begin{aligned} V^{\pi}(s) &= E_{s_t \sim T}\Big[\sum_{t=0}^{\infty} \gamma^t R(s_t, \pi(s_t)) | \pi, s_0 = s\Big] \\ &= E_{s_t \sim T}\Big[R(s_0, \pi(s_0)) + \sum_{t=1}^{\infty} \gamma^t R(s_t, \pi(s_t)) | \pi, s_0 = s\Big] \\ &= R(s, \pi(s)) + E_{s_t \sim T}\Big[\sum_{t=1}^{\infty} \gamma^t R(s_t, \pi(s_t)) | \pi\Big] \end{aligned}$$

$$= R(s, \pi(s)) + \gamma E_{s_t \sim T}\Bigg[\sum_{t=0}^{\infty} \gamma^t R(s_t, \pi(s_t)) | \pi, s_0 \sim T\Bigg]$$

$$= R(s, \pi(s)) + \gamma \sum_{s' \in S} T(s, \pi(s), s') V^{\pi}(s').$$

The last equation is known as Bellman equation which recursively finds the value function, defined as:

$$V^{\pi}(s) = \Bigg[R(s, \pi(s)) + \gamma \sum_{s' \in S} T(s, \pi(s), s') V^{\pi}(s')\Bigg]. \tag{3.1}$$

And the optimal state-value function $V^*$ can be found by:

$$V^*(s) = \max_{\pi} V^{\pi}(s)$$

$$= \max_{\pi}\Bigg[R(s, \pi(s)) + \gamma \sum_{s' \in S} T(s, \pi(s), s') V^{\pi}(s')\Bigg].$$

We can also define Bellman value function as a function of state and action, $Q^{\pi}(s, a)$, which estimates the expected return of taking action $a$ in a given state $s$ and policy $\pi$:

$$Q^{\pi}(s, a) = \Bigg[R(s, a) + \gamma \sum_{s' \in S} T(s, a, s') V^{\pi}(s')\Bigg]. \tag{3.2}$$

### 3.1.2 Partially Observable Markov Decision Processes

A POMDP is a more generalized framework for planning under uncertainty where the basic assumption is that the states are only partially observable. A POMDP is represented as a tuple $(S, A, T, \gamma, R, O, \Omega, b_0\}$. That is, a POMDP model includes an MDP model and adds:

- $O$ is the set of observations,
- $\Omega$ is the observation model:

$$\Omega(a, s', o') = \Pr(o' | a, s'),$$

  for the probability of observing $o'$ after taking the action $a$ which resulted in the state $s'$,
- and $b_0$ is an initial belief over all states.

Since POMDPs consider the environment partially observable, in POMDPs a belief over states is maintained in the run time as opposed to MDPs which consider states fully observable. So, in the run time if the POMDP belief over state $s$ at the current time is $b(s)$, then after taking action $a$ and observing observation $o$ the POMDP belief in the next time for state $s'$ is denoted by $b'(s')$ and is updated using the *State Estimator* function $\text{SE}(b, a, o')$:

$$\begin{aligned} b'(s') &= \text{SE}(b, a, o') \\ &= \Pr(s'|b, a, o') \\ &= \eta \Omega(a, s', o') \sum_{s \in S} b(s) T(s, a, s'), \end{aligned} \tag{3.3}$$

where $\eta$ is the normalization factor, defined as:

$$\eta = \frac{1}{\Pr(o'|b, a)}$$

and

$$\Pr(o'|b, a) = \sum_{s' \in S} \left[ \Omega(a, s', o') \sum_{s \in S} b(s) T(s, a, s') \right],$$

that is the probability of observing $o'$ after performing action $a$ in the belief $b$.

The reward function can also be defined on the beliefs:

$$R(b, a) = \sum_{s \in S} b(s) R(s, a). \tag{3.4}$$

Note, an important property of the belief state is that it is a sufficient statistics. In words, the belief at time $t$, i.e., $b_t$, summarizes the initial belief $b_0$, as well as all the actions taken and all observation received (Kaelbling et al. 1998). Formally, we have:

$$b_t(s) = \Pr(s|b_0, a_0, o_0, \ldots, a_{t-1}, o_{t-1}).$$

The POMDP policy selects an action $a$ for a belief state $b$, i.e., $a = \pi(b)$. In the POMDP framework the objective is to find an optimal policy $\pi^*$, where for any belief $b$, $\pi^*$ specifies an action $a = \pi^*(b)$ that maximizes the expected discount of future rewards starting from belief $b_0$:

$$\begin{aligned} V^{\pi}(b) &= E_{b_t \sim \text{SE}} \left[ \gamma^0 R(b_0, \pi(b_0)) + \gamma^1 R(b_1, \pi(b_1)) + \cdots | \pi, b_0 = b \right] \\ &= E_{b_t \sim \text{SE}} \left[ \sum_{t=0}^{\infty} \gamma^t R(b_t, \pi(b_t)) | \pi, b_0 = b \right]. \end{aligned}$$

Similar to MDPs, the value function of a policy can also be recursively defined as:

$$\begin{aligned}
V^{\pi}(b) &= E_{b_t \sim \mathrm{SE}}\Big[\gamma^0 R(b_0, \pi(b_0)) + \gamma^1 R(b_1, \pi(b_1) + \cdots) | \pi, b_0 = b\Big] \\
&= E_{b_t \sim \mathrm{SE}}\Big[\sum_{t=0}^{\infty} \gamma^t R(b_t, \pi(b_t)) | \pi, b_0 = b\Big] \\
&= R(b, \pi(b)) + E_{s_t \sim \mathrm{SE}}\Big[\sum_{t=1}^{\infty} \gamma^t R(b_t, \pi(b_t)) | \pi\Big] \\
&= R(b, \pi(b)) + \gamma E_{b_t \sim \mathrm{SE}}\Big[\sum_{t=0}^{\infty} \gamma^t R(b_t, \pi(b_t)) | \pi, b_0 \sim \mathrm{SE}\Big] \\
&= R(b, \pi(b)) + \gamma \sum_{o' \in O} \Pr(o' | b, \pi(b)) V^{\pi}(b').
\end{aligned}$$

The last equation is Bellman equation for POMDPs, defined as:

$$V^{\pi}(b) = \Big[R(b, \pi(b)) + \gamma \sum_{o' \in O} \Pr(o' | b, \pi(b)) V^{\pi}(b')\Big]. \tag{3.5}$$

Then, we have the optimal policy $\pi^*$ as:

$$\pi^*(b) = \operatorname*{argmax}_{\pi} V^{\pi}(b).$$

And the optimal belief-value model $V^*$ can be found by:

$$\begin{aligned}
V^*(b) &= \max_{\pi} V^{\pi}(b) \\
&= \max_{\pi}\Big[R(b, \pi(b)) + \gamma \sum_{o' \in O} \Pr(o' | b, \pi(b)) V^{\pi}(b')\Big].
\end{aligned}$$

We can also define Bellman value function as a function of beliefs and actions, $Q^{\pi}(b, a)$, which estimates the expected return of taking action $a$ in a given belief $b$ and policy $\pi$:

$$Q^{\pi}(b, a) = R(b, a) + \gamma \sum_{o' \in O} \Pr(o' | a, b) V^{\pi}(b'),$$

where $b' = \mathrm{SE}(b, a, o')$, is calculated from Eq. (3.3).

Notice that we can see a POMDP as an MDP, if the POMDP includes a deterministic observation model and a deterministic initial belief. This can be seen

**Table 3.1** The process of policy learning in the $Q$-learning algorithm (Schatzmann et al. 2006)

| | $s_1$ | $s_2$ | $s_3$ | $s_4$ | $s_5$ | ... |
|---|---|---|---|---|---|---|
| $a_1$ | 4.23 | 5.67 | 2.34 | 0.67 | **9.24** | ... |
| $a_2$ | 1.56 | **9.45** | **8.82** | 5.81 | 2.36 | ... |
| $a_3$ | 4.77 | 3.39 | 2.01 | **7.58** | 3.93 | ... |
| ... | ... | ... | ... | ... | ... | ... |

in Eq. (3.3), by starting with a deterministic initial belief, the next belief will be deterministic as the observation model is deterministic. This means that such a POMDP knows its current state with 100 % probability similar to MDPs.

### 3.1.3 Reinforcement Learning

In Sect. 3.1, we introduced model-free RL, in short RL, which is performed when the environment model is not known. An algorithm known as $Q$-learning (Watkins and Dayan 1992) can be used for RL. These values estimate the expected return of taking action $a$ in state $s$ and following thereafter, as expressed in Eq. (3.2). The process of policy learning in the $Q$-learning algorithm can be seen in the matrix of Table 3.1, taken from Schatzmann et al. (2006). The $Q$-values are initialized with an arbitrary value for every pair $(s, a)$. The $Q$-values are iteratively updated to become better estimates of the expected return of the state-action pairs. While the agent is interacting with the environment the $Q$-values are updated using:

$$Q(s, a) \leftarrow (1 - \alpha)Q(s, a) + \alpha( R(s, a) + \gamma \max_{a'} Q(s', a')),$$

where $\alpha$ represents a learning rate parameter that decays from 1 to 0. When the $Q$-values for each state action pair is estimated, the optimal policy for each state selects the action with the highest expected value, i.e., the bolded values in Table 3.1.

In this book, our focus is on learning the dialog MDP/POMDP model components and then solve the dialog MDP/POMDP using the available planning algorithms. As such, we study the planning algorithms for solving MDPs/POMDPs in the following section.

### 3.1.4 Solving MDPs/POMDPs

Solving MDPs/POMDPs can be performed when the model components of the MDP or POMDP are defined/learned in advance. That is, solving the underlying MDP/POMDP for a near optimal policy. This is done by applying various model-based algorithms which work using dynamic programming (Bellman 1957a). Such algorithms fall into two categories of policy iteration and value iteration (Sutton and Barto 1998). In the rest of this section, we describe the policy iteration and value iteration for the MDP framework, respectively, in Sect. 3.1.4.1 and in Sect. 3.1.4.2. Then in Sect. 3.1.4.3, we introduce the value iteration for the POMDP framework. Since the value iteration algorithm for POMDPs is intractable, we study an approximated value iteration algorithm for the POMDP framework, known as point-based value iteration (PBVI) in Sect. 3.1.4.4.

#### 3.1.4.1 Policy Iteration for MDPs

Policy iteration methods have a general way of solving the value function in MDPs. They find the optimal value function by iterating on two phases known as *policy evaluation* and *policy improvement* shown in Algorithm 2. In Line 3, a random policy is selected, i.e., the policy $\pi_t$ is randomly initialized at $t = 0$. Then a random subsequent value of the policy is selected, i.e., the value $V_k$ is randomly chosen when $k = 0$. The algorithm then iterates on the two steps of policy evaluation and policy improvement.

In the policy evaluation step, i.e., Line 7, the algorithm calculates the value of policy $\pi_{t+1}$. This is done efficiently by calculating the value of $V_{k+1}$ using the value function $V_k$ of previous policy $\pi_t$, and then repeating this calculation until it finds a converged value for $V_k$. This is formally done as follows:

**Algorithm 2:** The policy iteration algorithm for MDPs

**Input**: An MDP model $\langle S, A, T, R\rangle$ ;
**Output**: A (near) optimal policy $\pi^*$;

`/* Initialization */`
1 $t \leftarrow 0$;
2 $k \leftarrow 0$;
3 $\forall s \in S$: Initialize $\pi_t(s)$ with an arbitrary action;
4 $\forall s \in S$: Initialize $V_k(s)$ with an arbitrary value;
5 **repeat**
`/* Policy evaluation */`
6 **repeat**
7 $\forall s \in S : V_{k+1}(s) \leftarrow R(s, \pi_t(s)) + \gamma \sum_{s' \in S} T(s, \pi_t(s), s')V_k(s')$;
8 $k \leftarrow k + 1$;
9 **until** $\forall s \in S : |V_k(s) - V_{k-1}(s)| < \epsilon$;
`/* Policy improvement */`
10 $\forall s \in S : \pi_{t+1}(s) \leftarrow \arg\max_{a \in A} \Big[ R(s, a) + \gamma \sum_{s' \in S} T(s, a, s')V_k(s') \Big]$;
11 $t \leftarrow t + 1$;
12 **until** $\pi_t = \pi_{t-1}$;
13 $\pi^* = \pi_t$;

$$\forall s \in S : V_{k+1}(s) \leftarrow R(s, \pi_t(s)) + \gamma \sum_{s' \in S} T(s, \pi_t(s), s') V_k(s').$$

The algorithm iterates until for all states $s$ the state values stabilize. That is, we have: $|V_k(s) - V_{k-1}(s)| < \epsilon$, where $\epsilon$ is a predefined threshold for error.

Then, in the policy improvement step, i.e., Line 10, the *greedy* policy $\pi_{t+1}$ is chosen. Formally, given the value function $V_k$, we have:

$$\forall s \in S : \pi_{t+1}(s) \leftarrow \arg\max_{a \in A} \left[ R(s, a) + \gamma \sum_{s' \in S} T(s, a, s') V_k(s') \right].$$

The process of policy evaluation and policy improvement continues until $\pi_t = \pi_{t+1}$. Then, policy $\pi_t$ is the optimal policy, i.e., $\pi_t = \pi^*$.

The significant drawback of the policy iteration algorithms is that for each improved policy $\pi_t$, a complete policy evaluation is done (Lines 7 and 8). Generally, value iteration algorithm is used to handle this drawback. We study value iteration algorithms for both MDPs and POMDPs in the following sections.

#### 3.1.4.2 Value Iteration for MDPs

Value iteration methods overlap the evaluation and improvement steps introduced in the previous section. Algorithm 3 demonstrates the value iteration method in MDPs. It consists of a backup operation as:

$$\forall s \in S : V_{k+1}(s) \leftarrow \max_{a \in A} \left[ R(s, a) + \gamma \sum_{s' \in S} T(s, a, s') V_k(s') \right].$$

---

**Algorithm 3:** The value iteration algorithm for MDPs

**Input**: An MDP model $\langle S, A, T, R \rangle$ ;
**Output**: A (near) optimal policy $\pi^*$;

$k \leftarrow 0$;
$\forall s \in S$: Initialize $V_k(s)$ with an arbitrary value;
**repeat**
    $\forall s \in S : V_{k+1}(s) \leftarrow \max_{a \in A} \left[ R(s, a) + \gamma \sum_{s' \in S} T(s, a, s') V_k(s') \right]$;
    $k \leftarrow k + 1$;
**until** $\forall s \in S : |V_k(s) - V_{k-1}(s)| < \epsilon$;
$\forall s \in S : \pi^*(s) \leftarrow \arg\max_{a \in A} \left[ R(s, a) + \gamma \sum_{s' \in S} T(s, a, s') V_k(s') \right]$;

---

This operation continues in Lines 4 and 5 until for all states $s$, state values stabilize. That is, we have: $|V_k(s) - V_{k-1}(s)| < \epsilon$. Then, the optimal policy is the greedy policy with regard to the value function shown in Line 4.

#### 3.1.4.3 Value Iteration for POMDPs

Solving POMDPs is more challenging than solving MDPs. To find the solution of a MDP, an algorithm such as value iteration needs to find the optimal policy for $|S|$ discrete states. However, finding the solution of POMDPs is more challenging, since the algorithm, such as value iteration, needs to find the solution for $|S| - 1$ dimensional *continuous* space. This problem is called *curse of dimensionality* in POMDPs (Kaelbling et al. 1998). Then, the POMDP solution is found as a breadth first search in $t$-steps, for the beliefs that have been created in the $t$-steps. This is called $t$-step *planning*. Notice that the number of created beliefs increases exponentially with respect to the planning time $t$. This problem is called *curse of history* in POMDPs (Kaelbling et al. 1998; Pineau 2004).

Planning is performed in POMDPs as a breadth first search in trees for a finite $t$, and consequently finite $t$-step conditional plans. A $t$-step conditional plan describes a policy with a horizon of $t$-step further (Williams 2006). It can be represented as a tree that includes a specified root action $a_t$. Figure 3.2 shows a 3-step conditional plan in which the root is indexed with time step $t$ ($t = 3$) and the leafs are indexed with time step 1. The edges are indexed with observations that lead to a node at $t - 1$ level, representing a $t - 1$-step conditional plan.

Each $t$-step conditional plan has a specific value $V_t(s)$ for unobserved state $s$ which is calculated as:

$$V_t(s) = \begin{cases} 0 & \text{if } t = 0; \\ R(s, a_t) + \gamma \sum_{s' \in S} T(s, a_t, s') \sum_{o' \in O} \Omega(a_t, s', o') V_{t-1}^{o'}(s') & \text{otherwise}; \end{cases}$$

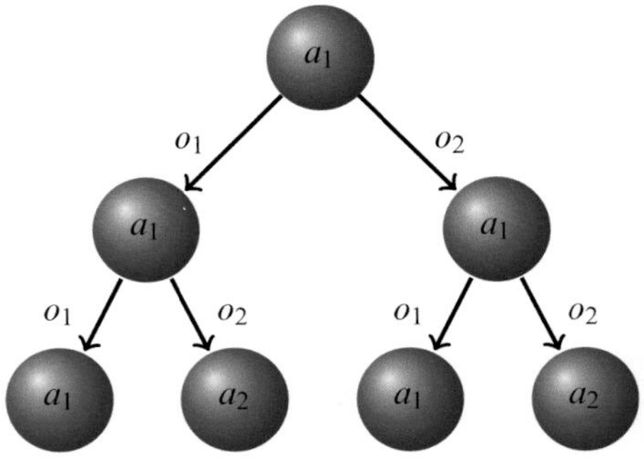


**Fig. 3.2** A 3-step conditional plan of a POMDP with two actions and two observations. Each node is labeled with an action and each non-leaf node has exactly $|O|$ observations

where $a_t$ is the specified action for the $t$-step conditional plan. Moreover, $V_{t-1}^{o'}(s')$ is the value of $t-1$-step conditional plan (in level $t-1$) which is the child index $o'$ of $t$ conditional plan (with root node $a_t$).

Since in POMDPs the state is unobserved and a belief over possible states is maintained then the value of $t$-step conditional plan is calculated in runtime using the current belief $b$. More specifically, the value of $t$-step conditional plan for belief $b$, denoted by $V_t(b)$, is an expectation over states:

$$V_t(b) = \sum_{s \in S} b(s) V_t(s).$$

In POMDPs, given a set of $t$-step conditional plans, the agent's task is to find the conditional plan that maximizes the belief's value. Formally, given a set of $t$-step conditional plans denoted by $N_t$, in which the plans' indices are denoted by $n$, the best $t$-step conditional plan is the one that maximizes the belief's value:

$$V_t^*(b) = \max_{n \in N_t} \sum_{s \in S} b(s) V_t^n(s), \tag{3.6}$$

where $V_t^n$ is the $n$th $t$-step conditional plan.

And, the optimal policy for belief $b$ is calculated as:

$$\pi^*(b) = a_t^n,$$

where $n = \arg\max_{n \in N_t} \sum_{s \in S} b(s) V_t^n(s)$.

The value of each $t$-step conditional plan, $V_t(b)$, is a hyperplane in belief state, since it is an expectation over states. Moreover, the optimal policy takes the max over many hyperplanes, this causes the value function, Eq. (3.6), to be piece-wise-linear and convex. The optimal value function is then formed of regions where one hyperplane (one conditional plan) is optimal (Sondik 1971; Smallwood and Sondik 1973).

After this introduction of planning for POMDPs, now we can go through value iteration in POMDPs. Algorithm 4, adapted from Williams (2006), describes value iteration for POMDPs (Monahan 1982; Kaelbling et al. 1998). Value iteration proceeds by finding the subset of possible $t$-step conditional plans which contribute to the optimal $t$-step policy. These conditional plans are called *useful*, and only useful $t$-step plans are considered when finding the $(t+1)$-step optimal policy. In this algorithm, the input is a POMDP model and the planning time *maxT*, and the output is the set of *maxT*-step conditional plans, denoted by $V_{maxT}^n$, and their subsequent actions, denoted by $a_{maxT}^n$.

Each iteration of the algorithm contains two steps of generation and pruning. In the generation steps, Lines 4–11, the possibly useful $t$-step conditional plans are generated by enumerating all actions followed by all possible useful combinations of $(t-1)$-step conditional plans. This is done in Line 8:

---

**Algorithm 4:** The value iteration algorithm in POMDPs adapted from Williams (2006)

---

**Input**: A POMDP model $\langle S, A, T, \gamma, R, O, \Omega, b_0 \rangle$ and $maxT$ for planning horizon;
**Output**: The conditional plan $V_{maxT}^n$ and its subsequent action $a_{maxT}^n$;

1  $\forall s \in S$: Initialize $V_0(s)$ with 0 ;
2  $N \leftarrow 1$;
   /* N is the number of $t-1$ step conditional plans                    */
3  **for** $t \leftarrow 1$ *to maxT* **do**
     /* Generate $\{v^{a,k}\}$, the set of possibly useful conditional
     plans                                                        */
4      $K \leftarrow \{V_{t-1}^n : 1 \leq n \leq N\}^{|O|}$ ;
     /* $K$ now contains $N^{|O|}$ elements, where each element $k$ is a
     vector $k = (V_{t-1}^{x_1}, \ldots, V_{t-1}^{x_{|O|}})$. This growth is the source of the
     computational complexity                                      */
5      **foreach** $a \in A$ **do**
6         **foreach** $k \in K$ **do**
7            **foreach** $s \in S$ **do**
            /* Notation $k(o')$ refers to element $o'$ of vector $k$.
            */
8              $v^{a,k}(s) \leftarrow R(s,a) + \gamma \sum_{s' \in S} \sum_{o' \in O} T(s,a,s')\Omega(a,s',o')V_{t-1}^{k(o')}(s')$;
9            **end**
10        **end**
11     **end**
     /* Prune $\{v^{a,k}\}$ to yield $\{V_t^n\}$, set of actually useful CPs    */
     /* $n$ is the number of $t$-step conditional plans               */
12     $n \leftarrow 0$;
13     **foreach** $a \in A$ **do**
14        **foreach** $k \in K$ **do**
15           // If the value of plan $v^{a,k}$ is optimal in any belief, it is *useful* and will be kept.;
16           **if** $\exists b : v^{a,k}(b) = \max_{\bar{a},\bar{k}} v^{\bar{a},\bar{k}}(b)$ **then**
17             $n \leftarrow n + 1$;
18             $a_t^n \leftarrow a$;
19             **foreach** $s \in S$ **do**
20                $V_t^n(s) \leftarrow v^{a,k}(s)$;
21             **end**
22           **end**
23        **end**
24     **end**
25     $N \leftarrow n$;
26 **end**

---

$$v^{a,k} \leftarrow R(s,a) + \gamma \sum_{s' \in S} \sum_{o' \in O} T(s,a,s')\Omega(a,s',o')V_{t-1}^{k(o')},$$

where $k(o')$ refers to element $o'$ of the vector $k = (V_{t-1}^{n_1}, \ldots, V_{t-1}^{n_{|O|}})$.

Then, pruning is done in Lines 12–25. In the pruning step, the conditional plans that are not used in the optimal $t$-step policy are removed, which remains the set of useful $t$-step conditional plans. In particular, in Line 16, if there is a belief where $v^{a,k}$ makes the optimal policy, then the $n$th index of $t$-step conditional plan is set to $v^{a,k}$, i.e., $V_t^n(s) = v^{a,k}$.

Notice that value iteration for POMDPs is exponential to the number of observations (Cassandra et al. 1995). In fact, it has been proved that finding the optimal policy of a POMDP is a PSPACE-complete problem (Papadimitriou and Tsitsiklis 1987; Madani et al. 1999). Even finding a near optimal policy, i.e., a policy with a bounded value loss compared to the optimal one is NP-hard for a POMDP (Lusena et al. 2001).

As introduced in the beginning of this section, the main challenge for planning in POMDPs is because of curse of dimensionality and curse of history. So, numerous approximate algorithms for planning in POMDPs have been proposed in the past. For instance, Smallwood and Sondik (1973) developed a variation of value iteration algorithm for POMDPs. Other approaches include point-based algorithms (Pineau et al. 2003; Pineau 2004; Smith and Simmons 2004; Spaan and Spaan 2004; Paquet et al. 2005), heuristic-based method of Hauskrecht (2000), structure-based algorithms (Bonet and Geffner 2003; Dai and Goldsmith 2007; Dibangoye et al. 2009), compression-based algorithms (Lee and Seung 2001; Roy et al. 2005; Poupart and Boutilier 2002; Li et al. 2007), and forward search algorithms (Paquet 2006; Ross et al. 2008). In this context, the PBVI algorithms (Pineau et al. 2003) perform the planning for a fixed set of belief points. In the following section, we study the PBVI algorithm as described in Williams (2006).

#### 3.1.4.4 PBVI for POMDPs

Value iteration for POMDPs is computationally complex, because it tries to find an optimal policy for all belief points in the belief space. As such, not all of the generated conditional plans (in the generation step of value iteration) can be processed in the pruning step. In fact, in the pruning step there is a search for a belief in continuously valued space of beliefs (Williams 2006). On the other hand, the PBVI algorithm (Pineau et al. 2003) works by searching optimal conditional plans only at a finite set of $N$ discrete belief points $\{b_1, \ldots, b_N\}$. That is, each unpruned conditional plan $V_t^n(s)$ is exact only at belief $b_n$, and consequently PBVI algorithms are approximate planning algorithms for POMDPs.[1]

Algorithm 5 adapted from Williams (2006) describes the PBVI algorithm. The input and output of the algorithm is similar to the value iteration algorithm for POMDPs. Here, the input adds a set of $N$ random discrete belief points (besides the POMDP model and the planning time *maxT* which is used also in value iteration

[1]Note that here we assume that the PBVI is performed on a fixed set of random points similar to the PERSEUS algorithm, the PBVI algorithm proposed by Spaan and Vlassis (2005).

**Algorithm 5:** Point-based value iteration algorithm for POMDPs adapted from Williams (2006)

**Input**: A POMDP model $\langle S, A, T, \gamma, R, O, \Omega, b_0 \rangle$, $maxT$ for planning horizon, and a set of $N$ random beliefs $B$;
**Output**: The conditional plan $V^n_{maxT}$ and its subsequent action $a^n_{maxT}$;

1 **for** $n \leftarrow 1$ *to* $N$ **do**
2     **foreach** $s \in S$ **do**
3         $V^n_0(s) \leftarrow 0$;
4     **end**
5 **end**
6 **for** $t \leftarrow 1$ *to* $T$ **do**
    `/*` Generate $\{v^{a,k}\}$, the set of possibly useful conditional plans `*/`
7     **for** $n \leftarrow 1$ *to* $N$ **do**
8         **foreach** $a \in A$ **do**
9             **foreach** $o' \in O$ **do**
10                 $b^{a,o'}_n \leftarrow SE(b_n, a, o')$;
11                 $m(o') \leftarrow \arg\max_{n_i} \sum_{s' \in S} b^{a,o'}_n(s') V^{n_i}_{t-1}(s')$;
12             **end**
13             **foreach** $s \in S$ **do**
14                 $v^{a,n}(s) \leftarrow R(s,a) + \gamma \sum_{s' \in S} \sum_{o' \in O} T(s,a,s') \Omega(a,s',o') V^{m(o')}_{t-1}(s')$;
15             **end**
16         **end**
17     **end**
    `/*` Prune $\{v^{a,n}\}$ to yield $\{V^n_t\}$, set of actually useful CPs `*/`
18     **for** $n \leftarrow 1$ *to* $N$ **do**
19         $a^t_n \leftarrow \arg\max_a \sum_{s \in S} b_n(s) v^{a,n}(s)$;
20         **foreach** $s \in S$ **do**
21             $V^n_t(s) \leftarrow v^{a^n_t,n}(s)$;
22         **end**
23     **end**
24 **end**

for POMDPs). And, the output is the set of $maxT$-step conditional plans, denoted by $V^n_{maxT}$, and their subsequent actions, denoted by $a^n_{maxT}$.

Similar to value iteration for POMDPs, the PBVI algorithm consists of two steps of generation and pruning. In Lines 7–17, the possibly useful $t$-step conditional plans are generated using the $N$ given belief points to the algorithm. First, for each given belief point, the next belief is formed for all possible action observation pairs; denoted by $b^{a,o'}_n$ in Line 10. Then, for each updated belief, $b^{a,o'}_n$, the index of the best $t-1$-step conditional plan is stored; denoted by $m(o')$ in Line 11. That is, the $t-1$-step conditional plan that brings the highest value for the updated belief, which is calculated as:

$$m(o') \leftarrow \arg\max_{n_i} \sum_{s' \in S} b^{a,o'}_n(s') V^{n_i}_{t-1}(s').$$

The final task in the generation step of PBVI is generating a set of possible useful conditional plan for the current belief and action, denoted by $\upsilon^{a,n}$ which is calculated for each state in Line 14 as:

$$\upsilon^{a,n}(s) \leftarrow R(s,a) + \gamma \sum_{s' \in S} \sum_{o' \in O} T(s,a,s') \Omega(a,s',o') V_{t-1}^{m(o')}(s'),$$

where $V_{t-1}^{m(o')}(s')$ is the best $t-1$-step conditional plan for the updated belief $b_n^{a,o'}$.

Finally, the pruning step is done in Lines 18–23. In the pruning step, for each given belief point $n$, the highest valued conditional plan is selected and the rest ones are pruned, in Line 19. This is done by finding the best action (the best $t$-step policy) from the generated conditional beliefs for the belief point $n$, i.e., $\upsilon^{a,n}$, which is calculated as:

$$a_n^t \leftarrow \arg\max_a \sum_{s \in S} b_n(s) \upsilon^{a,n}(s)$$

and its subsequent $t$-step conditional plan is stored as $V_t^n$ in Line 21.

In contrast to value iteration for POMDPs, the number of conditional plans are fixed in all iterations in the PBVI approach (which is equal to the number of the given belief points, $N$). This is because of the fact that each conditional plan is optimal at one of the belief points. Notice that although the set of found conditional plans are guaranteed to be optimal only at the finite set of given belief points, the hope is that they are optimal (or near optimal) for nearby belief points. Then, similar to value iteration the conditional plan for an arbitrary belief $b$ at run time is calculated using $\max_n b(s) V_t^n(s)$.

## 3.2 Spoken Dialog Management

The SDS of an intelligent machine is the system that is responsible for the interaction between machine and human users. Figure 3.3, adapted from Williams (2006), shows the architecture of an SDS. At the high level, an SDS consists of three modules: the input, the output, and the control. The input includes the automatic speech recognition (ASR) and natural language understanding (NLU) components. The output includes natural language generator (NLG) and text-to-speech (TTS) components. Finally, the control module is the core part of an SDS and consists of the dialog model and the dialog manager (DM). The control module is also called the dialog *agent* in this book.

The SDS modules work as follows. First, the ASR module receives the user utterance, i.e., a sequence of words in the form of speech signals, and makes an $N$-Best list containing all user utterance hypotheses. Next, NLU receives the noisy words from the ASR output, generates the possible intents that the user could have in mind, and sends them to the control module. The control module receives the

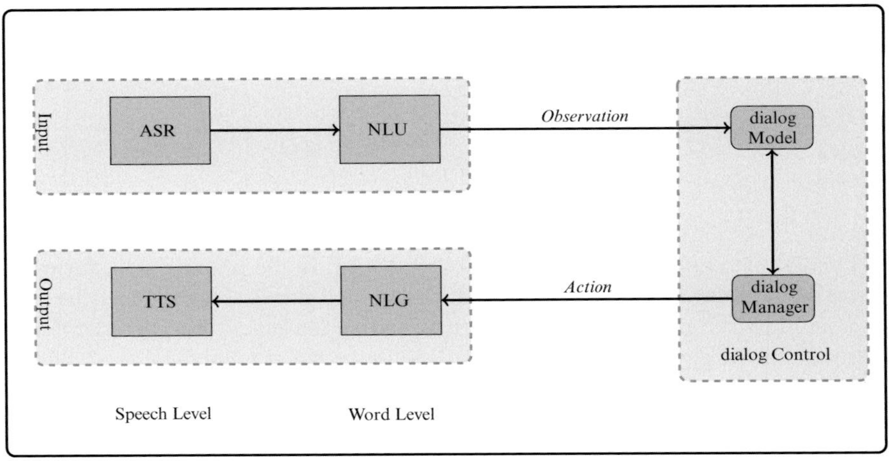


**Fig. 3.3** The architecture of a spoken dialog system, adapted from Williams (2006)

generated user intents, possibly with a confidence score, as an observation $O$. The confidence score can show, for instance, the reliability of possible user intents since the output generated by ASR and NLU can cause uncertainty in the machine. That is, the ASR output includes errors and the NLU output can be ambiguous, both cause uncertainty in SDS. The observation $O$ can be used in a dialog model to update and enhance the model.

Notice that the dialog model and the dialog manager interact with each other. In particular, the dialog model provides the dialog manager with the observation $O$ and the updated model. Based on such information, the dialog manager is responsible for making a decision. In fact, the DM updates its strategy based on the received updated model, and refers to its strategy for producing an action $A$, which is an input for NLG. The task of NLG is to produce a text describing the action $A$, and to pass the text to the TTS component. Finally, the TTS produces the spoken utterance of the text, and announces it for the user.

Note also that the dialog control part is the core part of an SDS, and is responsible for holding an efficient and natural communication with the user. To do so, the environment dynamics are approximated in the dialog model component over time. In fact, the dialog model aims to provide the dialog manager with better approximates of the environment dynamics. More importantly, the dialog manager is required to learn a strategy based on the updated model and to make a decision that satisfies the user intent during the dialog. But, this is a difficult task primarily because of the noisy ASR output, the NLU difficulties, and also the user intent change during the dialog. Thus, model learning and decision making is a significant task in SDS. In this context, the spoken dialog community modeled the dialog control of an SDS in the MDP/POMDP framework to automatically learn the dialog strategy, i.e., the dialog MDP/POMDP policy.

### 3.2.1 MDP-Based Dialog Policy Learning

In the previous section, we studied that the control module of an SDS is responsible for dialog modeling and management. The control module of a SDS, i.e., the dialog agent, has been formulated in the MDP framework so that the dialog MDP agent learns the dialog policy (Pieraccini et al. 1997; Levin and Pieraccini 1997). In this context, the MDP policy learning can be done either via model-free RL, or model-based RL. The model-free RL, in short RL, introduced in Sect. 3.1.3, can be done using techniques such as $Q$-learning. The model-based dialog policy learning is basically solving the dialog MDP/POMDP model using algorithms such as value iteration, introduced in Sect. 3.1.4.

In the model-based dialog policy learning, the dialog MDP model components can be given either by the domain experts manually, or learned from dialogs. In particular, the supervised learning approach can be used after annotating a dialog set to learn *user models*. For example, a user model can encode the probability of changing the user intent in each turn, given an executed machine's action. We study the user models further in Sect. 3.2.3. Then, the dialog MDP policy is learned using algorithms such as the value iteration algorithm, introduced in Sect. 3.1.4.2.

On the other hand, in the model-free RL which is also called simulation-based RL (Rieser and Lemon 2011), the dialog set is annotated and used for learning a *simulated environment*. Figure 3.4, taken from Rieser and Lemon (2011), shows a simulated environment. The dialog set is first annotated, and then used to learn the user model using supervised learning techniques. Moreover, the simulated environment requires an *error model*. The error model encodes the probability of

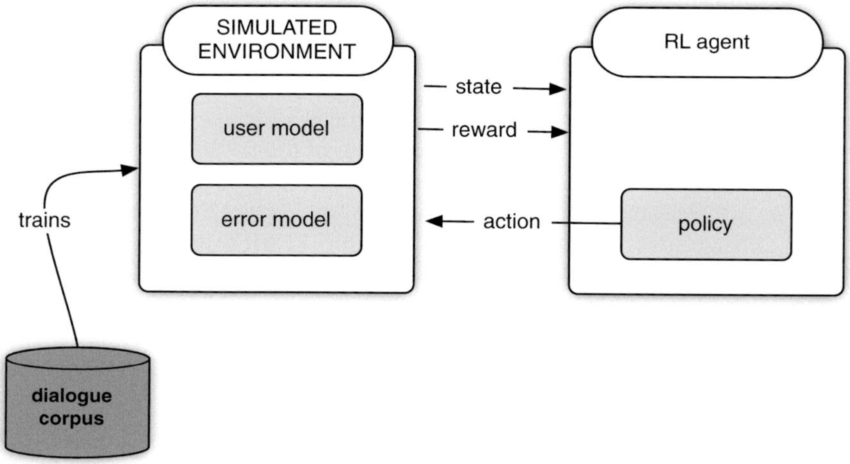


**Fig. 3.4** Simulation-based RL: learning a stochastic simulated dialog environment from data (Rieser and Lemon 2011)

occurring errors, for example by the ASR machine. The error model can be learned also from the dialog set. Then, model-free MDP policy learning techniques such as $Q$-learning (Sect. 3.1.3) are applied to learn the dialog MDP policy through interaction with the simulated user. For a comprehensive survey of recent advances in MDP-based dialog strategy learning (particularly simulation-based learning) the interested readers are referred to Frampton and Lemon (2009).

In contrast to MDPs, POMDPs are more general stochastic models that do not assume the environment's states fully observable, as introduced in Sect. 3.1. Instead, observations in POMDPs provide only partial information to the machine, and consequently, POMDPs maintain a belief over the states. As a result, the dialog POMDP policy performance is substantially higher than that of the dialog MDP policies, particularly in the noisy environments (Gašić et al. 2008; Thomson and Young 2010).

In this context, the POMDP-based dialog strategy learning is mostly model-based (Kim et al. 2011). This is mainly because reinforcement learning in POMDPs is a hard problem, and it is still being actively studied (Wierstra and Wiering 2004; Ross et al. 2008, 2011). In the next section, we present the related research on dialog POMDP policy learning.

### 3.2.2 POMDP-Based Dialog Policy Learning

The pioneer research for application of POMDPs in SDSs has been performed by Roy et al. (2000). The authors defined a dialog POMDP for SDS of a robot by considering possible user intents as the POMDP states. More specifically, their POMDP contained 13 states with mixture of six user intents and several user actions. In addition, the POMDP actions included ten clarifying questions as well as performance actions such as GOING TO A DIFFERENT ROOM, and PRESENTING INFORMATION TO USER.

For the choice of observations, the authors defined 15 *keywords* and an observation for the *nonsense* words. Moreover, the choice of the reward function has been hand-tuned. In fact, their defined reward function returned $-1$ for each dialog turn, that is for each clarification question regardless of the state of POMDP.

Then, Zhang et al. (2001b) proposed a dialog POMDP in the tourist guide domain. Their POMDP included 30 states with two factors, one factor with six possible user intents. The other factor encoded five values indicating the channel error such as *normal* and *noisy*. For the choice of the POMDP actions, the authors defined 18 actions such as ASKING USER'S INTENT and CONFIRMING USER'S INTENT.

Also, for the choice of the POMDP observations, Zhang et al. (2001b) defined 25 observations for the statement of user's intent, for instance *yes*, *no*, and *no response*. Moreover, for the reward function, they used a small negative reward for ASKING THE USER'S INTENT, a large positive reward for PRESENTING THE RIGHT INFORMATION FOR THE USER'S INTENT, and a large negative reward, otherwise. Finally, they used approximated methods to find their defined dialog POMDP solution and concluded that the POMDP approximate solution outperforms an MDP baseline.

Williams and Young (2007) also formulated the control module of SDSs in the POMDP framework. They factorized the machine's state to three components:

$$s = (g, u, d),$$

where $g$ is the user goal, which is similar to user intent, $u$ is the user action, i.e., the user utterance. In addition, $d$ is the dialog history, which indicates, for instance, what the user has said so far, or the user's view of what has been grounded in the conversation so far (Clark and Brennan 1991; Traum 1994). For a travel domain, the user goal could be any possible *(origin, destination)* pair allowed in the domain, for instance *(London, Edinburgh)*. Moreover, the user utterances could be similar to *from London to Edinburgh*. Finally, the machine's action could be such as WHICH ORIGIN and WHICH DESTINATION.

Williams and Young (2007) assumed that the user goal at each time step depends on the user goal and the machine's action in the previous time step:

$$\Pr(g'|g, a).$$

Moreover, they assumed that the user's action depends on the user goal and machine's action in the previous time step:

$$\Pr(u'|g', a).$$

Furthermore, the authors assumed that the current dialog history depends on the user goal and action, as well as the dialog history and the machine's action in the previous time step:

$$\Pr(d'|u', g', d, a).$$

Then, the state transition becomes:

$$\Pr(s'|s, a) = \underbrace{\Pr(g'|g, a)}_{\text{USER GOAL MODEL}} \cdot \underbrace{\Pr(u'|g', a)}_{\text{USER ACTION MODEL}} \cdot \underbrace{\Pr(d'|u', g', d, a)}_{\text{DIALOG HISTORY MODEL}} . \tag{3.7}$$

For the observation model, Williams and Young (2007) used the noisy recognized user's utterance $\tilde{u}$ together with confidence score $c$:

$$o = (u', c).$$

Moreover, they assumed that the machine's observation is based on the user's utterance and the confidence score $c$:

$$p(o'|s',a) = p(\tilde{u}',c'|u).$$

In addition, Williams and Young (2007) used a hand-coded reward function, for instance, large negative rewards for ASKING A NON-RELEVANT QUESTION, small negative reward for CONFIRMATION actions, and positive reward for ENDING the dialog successfully. In this way, the learned dialog POMDP policies try to minimize the number of turns and at the same time to finish a successful dialog.

Doshi and Roy (2007, 2008) proposed a dialog POMDP for a SDS of a robot. Similar to Roy et al. (2000), the authors considered the user's intent as POMDP states, for instance the user's intent for *coffee machine area*, or *main elevator*. In addition, they defined machine actions such as WHERE WOULD YOU LIKE TO GO, and WHAT WOULD YOU LIKE. Furthermore, the observations are the user utterances, for instance *I would like coffee*. In this work, the transition model encodes the probability of keywords given the machine's actions. For instance, given the machine's action WHERE DO YOU WANT TO GO, there is a high probability that the machine receives *coffee*, or *coffee machine*. Doshi and Roy (2008) used Dirichlet priors for uncertainty in the transition and observation models. In particular, for observation model they used Dirichlet counts and used an HMM to find the underlying states using EM algorithm.

Note that there are numerous other related works on dialog POMDPs. For instance, Doshi and Roy (2008) and Doshi-Velez et al. (2012) used active learning for learning dialog POMDPs. Thomson (2009), Thomson and Young (2010), Png and Pineau (2011), Atrash and Pineau (2010), and Png et al. (2012) used Bayesian techniques for learning dialog POMDP model components. In this context, Atrash and Pineau (2010) introduced a Bayesian method of learning an observation model for POMDPs which is explained further in Sect. 4.4. Moreover, Png and Pineau (2011) and Png et al. (2012) proposed an online Bayesian approach for updating the observation model of dialog POMDPs which is also described further in Sect. 4.4. On the other hand, Lison (2013) proposed a model-based Bayesian reinforcement learning approach to estimate the transition models for dialog management.

As mentioned, the learned dialog POMDP model components affect the optimized policy of the dialog POMDP. In particular, the transition model of a dialog POMDP usually includes the user model which needs to be learned from the dialog set. Kim et al. (2008) described different user model techniques that have been used in dialog POMDPs. These models are described in the following section.

### 3.2.3 *User Modeling in Dialog POMDPs*

In user modeling the goal is to model the distribution of plausible user responses, given a sequence of user acts and system responses, i.e., $\Pr(u_t|a_t, u_{t-1}, a_{t-1}, u_{t-2}, \ldots)$

from which an actual user response can be sampled (Young et al. 2013). One of the earliest approaches for user modeling is use of *n-grams* (Eckert et al. 1997; Georgila et al. 2005, 2006). The problem with these models is use of large $n$ to capture context. Also, the learned model is usually undertrained and thus it is difficult to ensure that the user model exhibited actions conform with actual user behavior.

To solve these problems, user models that are deterministic and goal oriented have been proposed with trainable random variables for when there is a user choice (Scheffler and Young 2000; Rieser and Lemon 2006; Pietquin 2006; Schatzmann et al. 2007; Ai and Litman 2007; Schatzmann and Young 2009; Keizer et al. 2010). Such user models yet need a lot of hand-tuning in order to work practically. Later on, dynamic Bayesian networks and hidden Markov models have been used for user modeling in POMDPs (Cuayáhuitl et al. 2005; Pietquin and Dutoit 2006).

The most recent approach for user modeling is use of POMDP-based dialog systems to behave like a user. A POMDP-based user (model) talks to the dialog (POMDP) system. The user POMDP and dialog POMDP, each has their policies for maximizing their reward functions. In this context, Chandramohan et al. (2011) used inverse reinforcement learning to infer users' reward functions (for the user POMDP) from real human–human dialogs.

In this section, we further describe the four user modeling techniques that have been used in dialog POMDPs (Kim et al. 2011). These models include *n-grams* (particularly the *bi-grams* and *tri-grams*) (Eckert et al. 1997), the *Levin* model (Levin and Pieraccini 1997), the *Pietquin* model (Pietquin 2004), and the *HMM* user model (Cuayáhuitl et al. 2005).

The bi-gram model learns the probability that the user performs action $u$, given the machine executes action $a$:

$$\Pr(u|a).$$

In tri-grams, the machine actions in two previous time-steps are considered. That is, the tri-gram model learns:

$$\Pr(u|a_n, a_{n-1}).$$

Thus, the Levin model reduces the number of parameters in the bi-grams by considering the *type* of the machine's action and learning the user actions for each type. These types include: *greeting, constraining*, and *relaxing* actions. The greeting action could be, for instance, HOW CAN I HELP YOU? The constraining actions are used to constrain a slot, for instance FROM WHICH CITY ARE YOU LEAVING? The relaxing actions are used for relaxing a constraint from a slot, for instance DO YOU HAVE OTHER DATES FOR LEAVING?

For the greeting action, the model learns:

$$\Pr(n),$$

where $n$ shows the number of slots for which the user provides info ($n = 0, 1, \ldots$). Also, the model learns the distribution on each slot $k$:

$$\Pr(k),$$

where $k$ is the slot number ($k = 1, 2, \ldots$).

For the constraining actions, the model learns two probability models. One is the probability that the user provides value for $n$ other slots while asked for slot $k$:

$$\Pr(n|k).$$

The other is the probability that the user provides value for slot $k'$ when it is asked for slot $k$:

$$\Pr(k'|k).$$

For the relaxing actions, the user either accepts the relaxation of the constraint or rejects it. So for each slot, the model learns:

$$\Pr(\text{yes}|k) = 1 - \Pr(\text{no}|k).$$

In the Levin model, however, the user goal is not considered in the user model. Then, the Pietquin model learns the probabilities conditioned on the user goal:

$$\Pr(u|a, g),$$

where $u$ is the user action (utterance), $g$ the user goal, and $a$ the machine's action. In this model the user goal is represented as a table of slot-value pairs. Since this can be a large table, an alternative approach can be considered. That is, for each part of the user goal, which is each slot, it is only maintained whether or not the user has provided information for that slot. So, for a dialog model with 4 slots, there exist only $2^4 = 16$ user goals. Note that in this way of user modeling the goal consistency is not maintained in the same way as the original Pietquin model.

In the HMM user modeling, first the probability of executing the machine's actions is learned based on the dialog state:

$$\Pr(a|d),$$

where $d$ is for the dialog state. Then, in the *input* HMM model, called IHMM, the model is enhanced by considering also the user actions besides the dialog state:

$$\Pr(a|d, u).$$

Finally, in the *input output* HMM, IOHMM, the user action model is learned based on the dialog state and the machine's action:

$$\Pr(u|d, a).$$

Note that in the above-mentioned works, the models are either assumed or have been learned from an annotated dialog set. In the following chapter, we propose methods for learning the dialog POMDP model components particularly the transition and observation models using unannotated dialogs and thus unsupervised learning techniques. Similar to Roy et al. (2000) and Doshi and Roy (2008), we use the user intents as POMDP states in this book. However, here we are interested in learning the dialog intents from the dialog set, rather than manually assigning them, and modeling the transition and observation models also based on unannotated dialogs.

# Chapter 4
# Learning the Dialog POMDP Model Components

## 4.1 Introduction

In this chapter, we propose methods for learning the model components of intent-based dialog POMDPs from unannotated and noisy dialogs. As stated in Chap. 1, in intent-based dialog domains, the dialog state is the user intent, where the users can mention their intents in different ways. In particular, we automatically learn the dialog states by learning the user intents from dialogs available in a domain of interest. We then learn a maximum likelihood transition model from the learned states. Furthermore, we propose two learned observation sets, and their subsequent observation models. The reward function however is learned in the next chapter where we present the IRL background and our proposed POMDP-IRL algorithms.

Note that we do not learn the discount factor since it is a number between 0 and 1 which is usually given. From the value function, shown in Eq. (3.5), we can see that if the discount factor is equal to 0, then the MDP/POMDP optimizes only immediate rewards, whereas if it is equal to 1, then the MDP/POMDP is in favor of future rewards (Sutton and Barto 1998). In SDS, for instance Kim et al. (2011) set the discount factor to 0.95 for all their experiments. We also hand tuned the discount factor to 0.90 for all our experiments. We set the initial belief state to the uniform distribution in all our experiments.

In the rest of this chapter, in Sect. 4.2, we learn the dialog POMDP states (Chinaei and Chaib-Draa 2014b). In this section, we first describe an unsupervised topic modeling approach known as hidden topic Markov model (HTMM) (Gruber et al. 2007); the method that we adapted for learning user intents from dialogs, in Sect. 4.2.1. We then present an illustrative example, using SACTI-1 dialogs (Williams and Young 2005), which shows the application of HTMM on dialogs for learning the user intents, in Sect. 4.2.2. We introduce our maximum likelihood transition model using the learned intents in Sect. 4.3. Then, we propose two observation sets and their subsequent observation models, learned from dialogs, in Sect. 4.4. We then revisit through the illustrative example on SACTI-1 to apply the proposed methods for



H. Chinaei, B. Chaib-draa, *Building Dialogue POMDPs from Expert Dialogues*,
SpringerBriefs in Speech technology, DOI 10.1007/978-3-319-26200-0_4

learning and training a dialog POMDP (without the reward function) in Sect. 4.5. In this section, we also evaluate the HTMM method for learning dialog intents, in Sect. 4.5.1, followed by the evaluation of the learned dialog POMDPs from SACTI-1 in Sect. 4.5.2. Finally, we conclude this chapter in Sect. 4.6.

## 4.2 Learning Intents as States

Recall our Algorithm 1, presented in Chap. 1, that shows the high level procedure for dialog POMDP model learning. The first step of the algorithm is to learn the states using an unsupervised learning method. As discussed earlier, the user intents are used as the dialog POMDP states. As such, in the first step we aim to capture the possible user intents in a dialog domain based on unannotated and noisy dialogs. Figure 4.1 represents dialog states as they are learned based on an unsupervised learning (UL) method. Here, we use HTMM (Gruber et al. 2007) to consider the Markovian property of states between $n$ and $n+1$ time steps. The HTMM method for intent learning from unannotated dialogs is as follows.

### *4.2.1 Hidden Topic Markov Model for Dialogs*

Hidden topic Markov model, in short HTMM (Gruber et al. 2007), is an unsupervised topic modeling technique that combines LDA (cf. Sect. 2.2) and HMM (cf. Sect. 2.3) to obtain the topics of documents. In Chinaei et al. (2009), we adapted HTMM for dialogs. A dialog set $\boldsymbol{D}$ consists of an arbitrary number of dialogs, $\boldsymbol{d}$. Similarly, each dialog $\boldsymbol{d}$ consists of the recognized user utterances, $\tilde{u}$, i.e., the ASR recognition of the actual user utterance $u$. The recognized user utterance, $\tilde{u}$, is a bag of words, $\tilde{u} = [w_1, \ldots, w_n]$.

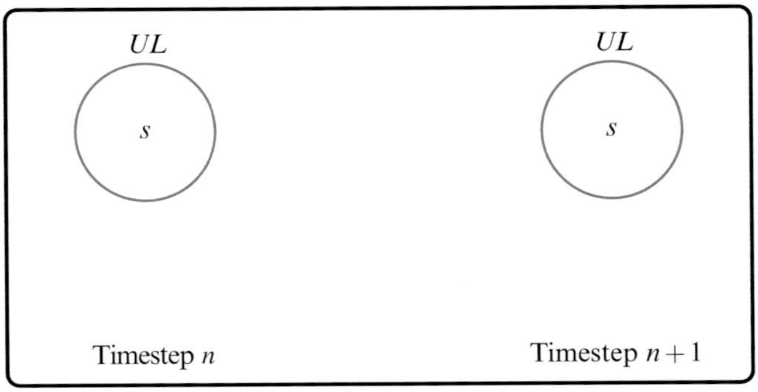


**Fig. 4.1** Hidden states are learned based on an unsupervised learning (UL) method that considers the Markovian property of states between $n$ and $n+1$ time steps. Hidden states are represented in the *light circles*

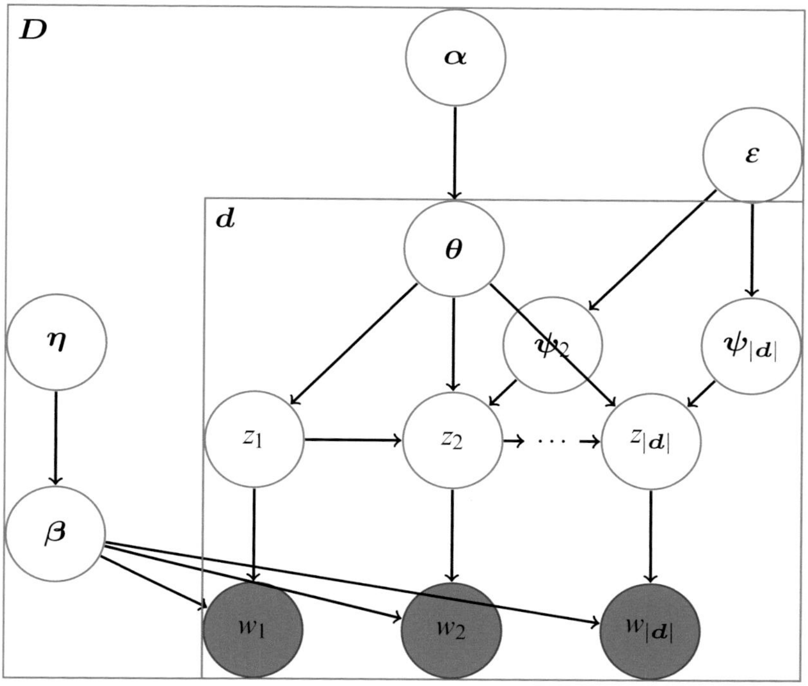


**Fig. 4.2** The HTMM model adapted from Gruber et al. (2007), the *shaded nodes* are words ($w$) used to capture intents ($z$)

Figure 4.2 shows the HTMM model, which is similar to the LDA model shown in Fig. 2.2. HTMM, however, applies the first-order Markov property to LDA, and is explained further in this section. Figure 4.2 shows that the dialog $\boldsymbol{d}$ in a dialog set $\boldsymbol{D}$ can be seen as a sequence of words $w_i$ which are observations for a hidden intents $z$. Since hidden intents are equivalent to user intents, hereafter, hidden intents are called user intents. The vector $\boldsymbol{\beta}$ is a global vector that ties all the dialogs in a dialog set $\boldsymbol{D}$, and retains the probability of words given user intents, $\Pr(w|z, \boldsymbol{\beta}) = \boldsymbol{\beta}_{wz}$. In particular, the vector $\boldsymbol{\beta}$ is drawn based on multinomial distributions with a Dirichlet prior $\boldsymbol{\eta}$. On the other hand, the vector $\boldsymbol{\theta}$ is a local vector for each dialog $\boldsymbol{d}$, and retains the probability of intents in a dialog, $\Pr(z|\boldsymbol{\theta}) = \boldsymbol{\theta}_z$. Moreover, the vector $\boldsymbol{\theta}$ is drawn based on multinomial distributions with a Dirichlet prior $\boldsymbol{\alpha}$.

The parameter $\boldsymbol{\psi}_i$ is for adding the Markovian property in dialogs since successive utterances are more likely to include the same user intent. The assumption here is that a recognized utterance represents only one user intent, so all the words in the recognized utterance are observations for the same user intent. To formalize that, the HTMM algorithm assigns $\boldsymbol{\psi}_i = 1$ for the first word of an utterance, and $\boldsymbol{\psi}_i = 0$ for the rest. Then, when $\boldsymbol{\psi}_i = 1$ (beginning of an utterance) a new intent is drawn, and when $\boldsymbol{\psi}_i = 0$ (in the utterance), the intent of the $n$th word is identical to the intent of

**Algorithm 6:** The HTMM generative model, adapted from Gruber et al. (2007)

**Input**: Set of dialogs $\boldsymbol{D}$, $N$ number of intents
**Output**: Generate utterances of $\boldsymbol{D}$

1 **foreach** *intent $z$ in the set of $N$ intents* **do**
2     Draw $\boldsymbol{\beta}_z \sim$ Dirichlet($\boldsymbol{\eta}$);
3 **end**
4 **foreach** *dialog $\boldsymbol{d}$ in $\boldsymbol{D}$* **do**
5     Draw $\boldsymbol{\theta} \sim$ Dirichlet($\boldsymbol{\alpha}$);
6     $\boldsymbol{\psi}_1 \leftarrow 1$;
7     **foreach** $i \leftarrow 2, \ldots, |\boldsymbol{d}|$ **do**
8         **if** *beginning of a user utterance* **then**
9             Draw $\boldsymbol{\psi}_i \sim$ Bernoulli($\epsilon$);
10         **else**
11             $\boldsymbol{\psi}_i \leftarrow 0$;
12         **end**
13     **end**
14     **foreach** $i \leftarrow 1, \ldots, |\boldsymbol{d}|$ **do**
15         **if** $\boldsymbol{\psi}_i = 0$ **then**
16             $z_i \leftarrow z_{i-1}$;
17         **else**
18             Draw $z_i \sim$ multinomial($\boldsymbol{\theta}$);
19         **end**
20         Draw $w_i \sim$ multinomial($\boldsymbol{\beta}_{z_i}$);
21     **end**
22 **end**

the previous one. Note that the parameter $\epsilon$ is used as a prior over $\boldsymbol{\psi}$ which controls the probability of intent transition between utterances in dialogs, $\Pr(z_i|z_{i-1}) = \epsilon$. Since each recognized utterance contains one user intent, we have $\Pr(z_i|z_{i-1}) = 1$ for $z_i, z_{i-1}$ within one utterance.

Algorithm 6 is the generative algorithm for HTMM, adapted from Gruber et al. (2007). This generative algorithm here is similar to the generative model of LDA introduced in Sect. 2.2. First, for all possible user intents, the vector $\boldsymbol{\beta}$ is drawn using the Dirichlet distribution with prior $\boldsymbol{\eta}$, in Line 2. Then, for each dialog, the vector $\boldsymbol{\theta}$ is drawn using the Dirichlet prior $\boldsymbol{\alpha}$. In Line 5, for each dialog, the vector $\boldsymbol{\theta}$ is initialized using the Dirichlet prior $\boldsymbol{\alpha}$.

In HTMM, however, for each recognized utterance $i$ in dialog $\boldsymbol{d}$, the parameter $\boldsymbol{\psi}$ is initialized based on a Bernoulli prior $\epsilon$ in Lines 7–13. As mentioned above, the parameter $\boldsymbol{\psi}$ basically adds the Markovian property to the model. It determines whether the user intent for the recognized utterance $i$ is the same as previous recognized utterance. The rest of the algorithm, Lines 14–21, finds the user intents. If the parameter $\boldsymbol{\psi}$ is equal to 0, the algorithm assumes that the user intent for utterance $i$ is equal to the one for utterance $i-1$, in Line 16, encoding thus the Markovian property. Otherwise, it draws the intent for utterance $i$ based on the vector $\boldsymbol{\theta}$ in Line 18. Finally, a new word $w$ is generated based on the vector $\boldsymbol{\beta}$, in Line 20.

HTMM uses Expectation Maximization (EM) and forward backward algorithm (Rabiner 1990) (cf. Sect. 2.3), the standard method for approximating the parameters in HMMs. This is due to the fact that conditioned on $\boldsymbol{\theta}$ and $\boldsymbol{\beta}$, HTMM is a special case of HMMs. In HTMM, the latent variables are user intents $z_i$ and $\boldsymbol{\psi}_i$ which determines if the intent for the word $w_i$ is drawn from $w_{i-1}$, i.e., if $\boldsymbol{\psi}_i = 0$; or a new intent will be generated, i.e., if $\boldsymbol{\psi}_i = 1$.

1. In the expectation step, the $Q$ function from Eq. (2.5) is instantiated. For each user intent $z$, we need to find the expected count of intent transitions to intent $z$.

$$E(C_{\boldsymbol{d},z}) = \sum_{j=1}^{|\boldsymbol{d}|} \Pr(z_{\boldsymbol{d},j} = z, \boldsymbol{\psi}_{\boldsymbol{d},j} = 1 | w_1, \ldots, w_{|\boldsymbol{d}_i|}),$$

where $\boldsymbol{d}$ is a dialog in the dialog set, $\boldsymbol{D}$.

Moreover, we need to find the expected number of co-occurrence of a word $w$ with an intent $z$.

$$E(C^{z,w}) = \sum_{i=1}^{|\boldsymbol{D}|} \sum_{j=1}^{|\boldsymbol{d}_i|} \Pr(z_{i,j} = z, w_{i,j} = w | w_1, \ldots, w_{|\boldsymbol{d}_i|}),$$

where $\boldsymbol{d}_i$ is the $i$th dialog in the dialog set $\boldsymbol{D}$, and $w_{i,j}$ is the $j$th word of the $i$th dialog.

2. In the maximization step, the maximum a posteriori (MAP) estimate for $\boldsymbol{\theta}$ and $\boldsymbol{\beta}$ is computed by the standard method of Lagrange multipliers (Bishop 2006):

$$\boldsymbol{\theta}_{\boldsymbol{d},z} \propto E(C_{\boldsymbol{d},z}) + \boldsymbol{\alpha} - 1$$

$$\boldsymbol{\beta}_{w,z} \propto E(C^{z,w}) + \boldsymbol{\eta} - 1.$$

Note that the vector $\boldsymbol{\theta}_z$ stores the probability of an intent $z$:

$$\Pr(z|\boldsymbol{\theta}) = \boldsymbol{\theta}_z. \tag{4.1}$$

And, the vector $\boldsymbol{\beta}_{w,z}$ stores the probability of an observation $w$ given the intent $z$:

$$\Pr(w|z, \boldsymbol{\beta}) = \boldsymbol{\beta}_{wz}. \tag{4.2}$$

The parameter $\epsilon$ denotes the dependency of the utterances on each other, i.e., how likely it is that two successive uttered utterances of the user have the same intent.

$$\epsilon = \frac{\sum_{i=1}^{|\boldsymbol{D}|} \sum_{j=1}^{|\boldsymbol{d}|} \Pr(\boldsymbol{\psi}_{i,j} = 1 | w_1, \ldots, w_{|\boldsymbol{d}|})}{\sum_{i=1}^{|\boldsymbol{D}|} N_{i,\text{utt}}},$$

where $N_{i,\text{utt}}$ is the number of utterances in the dialog $i$.

Learning the parameters in HTMM can be done in a small computation time, using EM. This is a useful property, though EM suffers from local minima (Ortiz and Kaelbling 1999) and the related work such as Griffiths and Steyvers (2004) proposed the Gibbs sampling method rather than EM. Ortiz and Kaelbling (1999), however, introduced methods for getting away from local minima, and also suggested that EM can be accelerated via some heuristics based on the type of the problem.

In HTMM, the special form of the transition matrix reduces the time complexity of the forward backward algorithm to $O(TN)$, where $T$ is the length of the chain, and $N$ is the number of desired user intents given to the algorithm (Gruber et al. 2007; Gruber and Popat 2007). The small computation time is particularly useful, as it allows the machine to update its model when it observes new data.

### 4.2.2 Learning Intents from SACTI-1 Dialogs

In this section, we apply HTMM on SACTI-1 dialogs (Williams and Young 2005), publicly available at: http://mi.eng.cam.ac.uk/projects/sacti/corpora/. SACTI stands for simulated ASR channel tourist information. It contains 144 dialogs between 36 users and 12 experts who play the role of the machine for 24 total tasks on this data set. The utterances are first recognized using a speech recognition error simulator, and then are sent to human experts for a response. There are four levels of ASR noise in SACTI-1 data: *none*, *low*, *medium*, and *high* noise. There is a total of 2048 utterances that we used for our experiments which have 817 distinct words.

Table 4.1 shows a dialog sample from SACTI-1. The first line of the table shows the first user utterance, $u_1$. Because of ASR errors, this utterance is recognized as $\tilde{u}_1$. Then, $m_1$ is the actual machine utterance as a response to the user request recognized by the ASR in $\tilde{u}_1$. We applied HTMM as introduced in the previous section to learn possible user intents in SACTI-1. In our experiments, we removed the machine responses from the dialogs in order to learn the user intents based on the recognized user utterances. Nevertheless, since HTMM is an unsupervised learning method, we did not have to annotate the dialogs.

Table 4.2 shows the learned intents from SACTI-1 data, using HTMM. The algorithm learns three user intents which we named them, respectively, as:

1. *visits*,
2. *transports*,
3. *foods*.

Each intent is represented by its 20-top words with their probabilities. In Table 4.2, we have highlighted only the words which best represents each intent. These highlighted words are called *keywords*. To extract keywords, we avoided stop words such as *the, a, an, to*. For instance, the words *hotel*, *tower*, and *castle* are keywords which represent the user intents for information necessary about visiting areas, i.e., *visits*.

**Table 4.1** A sample from the SACTI-1 dialogs (Williams and Young 2005)

| | |
|---|---|
| | ... |
| $u_1$ | yeah hello this is johan schmulka uh and i'm uh searching for a bar in this town can you may be tell me where the cafe blu is |
| $\tilde{u}_1$ | [hello this is now seven four bus and do you tell me where to cafe blu is] |
| $m_1$ | cafe blu is on alexander street |
| $u_2$ | oh um yeah how can i get to alexander street and where exactly is it i know there a shopping area on alexander street um |
| $\tilde{u}_2$ | [i am yeah i am at the alexander street and where is it was on a the center of alexander street] |
| $m_2$ | it is on the east side of alexander street so %um it's %um just off middle road |
| | ... |

Then, for each recognized user utterance $\tilde{u} = [w_1, \ldots, w_n]$, we define its subsequent state as the highest probable intent $z$:

$$s = \operatorname*{argmax}_z \Pr(w_1, \ldots, w_n|z)$$
$$= \operatorname*{argmax}_z \prod_i \Pr(w_i|z), \tag{4.3}$$

where $\Pr(w_i|z)$ is already learned and stored in the parameter $\boldsymbol{\beta}_{wz}$ according to Eq. (4.2). The second equality in the equation, the product of probabilities, is due to the independency of words given a user intent.

User intents have been previously suggested to be used as states of dialog POMDPs (Roy et al. 2000; Zhang et al. 2001b; Matsubara et al. 2002; Doshi and Roy 2007, 2008). However, to the best of our knowledge, they have not been automatically extracted from real data. Here, we learn the user intents based on unsupervised learning methods. This enables us to use raw data, with little annotation or preprocessing. In our previous work (Chinaei et al. 2009), we were able to learn 10 user intents from SACTI-2 dialogs (Weilhammer et al. 2004), without annotating data or any prepossessing. In this paper, we showed cases where we can estimate the user intents behind utterances when users did not use a keyword for an intent. In addition, we were able to learn the true intent behind recognized utterances that included wrong keywords or multiple keywords, possibly keywords of different learned intents.

We captured user intents based on unannotated expert dialogs in SACTI-2 dialog set (Weilhammer et al. 2004) using HTMM method introduced above (Chinaei et al. 2009). Afterwards, we used HTMM to learn user intents for another domain SACTI-1 (Williams and Young 2005), and used the user intents as the states of the dialog POMDP (Chinaei and Chaib-draa 2011).

**Table 4.2** The learned user intents from the SACTI-1 dialogs

| *intent* 1 | | *visits* | |
|---|---|---|---|
| the | 0.08 | like | 0.01 |
| i | 0.06 | **hotel** | 0.01 |
| to | 0.05 | for | 0.01 |
| um | 0.02 | would | 0.01 |
| is | 0.02 | i'm | 0.01 |
| a | 0.02 | **tower** | 0.01 |
| and | 0.02 | **castle** | 0.01 |
| you | 0.02 | go | 0.01 |
| uh | 0.02 | do | 0.01 |
| what | 0.01 | me | 0.01 |
| *intent* 2 | | *transports* | |
| the | 0.08 | a | 0.02 |
| to | 0.04 | does | 0.02 |
| is | 0.04 | **road** | 0.02 |
| how | 0.03 | and | 0.01 |
| um | 0.02 | on | 0.01 |
| it | 0.02 | long | 0.01 |
| uh | 0.02 | of | 0.01 |
| i | 0.02 | much | 0.01 |
| from | 0.02 | **bus** | 0.01 |
| **street** | 0.02 | there | 0.01 |
| *intent* 3 | | *foods* | |
| you | 0.06 | um | 0.02 |
| the | 0.04 | and | 0.20 |
| i | 0.04 | thank | 0.01 |
| a | 0.03 | to | 0.01 |
| me | 0.03 | of | 0.01 |
| is | 0.02 | **restaurant** | 0.01 |
| uh | 0.02 | there | 0.01 |
| can | 0.02 | do | 0.01 |
| tell | 0.02 | could | 0.01 |
| please | 0.02 | where | 0.01 |

## 4.3 Learning the Transition Model

In the previous section, we learned states of the dialog POMDP. In this section, we go through the second step of our descriptive Algorithm 1: extracting actions directly from dialogs and learning a maximum likelihood transition model.

In Sect. 3.1.1, we saw that a transition model is in the form of $T(s_1, a_1, s_2)$ where $T$ stores the probability of going to the state $s_2$ given performing the action $a_1$ in

**Table 4.3** Learned probabilities of intents for the recognized utterances in the SACTI-1 example

| | |
|---|---|
| | ... |
| $u_1$ | yeah hello this is johan schmulka uh and i'm uh searching for a bar in this town can you may be tell me where the cafe blu is |
| $\tilde{u}_1$ | [hello this is now seven four bus and do you tell me where to cafe blu is] |
| $\Pr_1$ | *t:0.00 v:0.00 f:1.00* |
| $u_2$ | oh um yeah how can i get to alexander street and where exactly is it i know there a shopping area on alexander street um |
| $\tilde{u}_2$ | [i am yeah i am at the alexander street and where is it was on a the center of alexander street] |
| $\Pr_2$ | *t:0.99 v:0.00 f:0.00* |
| | ... |

the state $s_1$. We learn a maximum likelihood transition model by performing the following counting:

$$T(s_1, a_1, s_2) = \Pr(z'|z, a) = \frac{\text{Count}(z_1, a_1, z_2)}{\text{Count}(z_1, a_1)}. \tag{4.4}$$

To do so, we extract the set of possible actions from the dialog set. Then, the maximum probable intent (state) is assigned to each recognized utterance using Eq. (4.3).

For instance, for the recognized utterances in the SACTI-1 example, we can learn the probability distribution of the intents from Eq. (4.2), denoted by $Pr$ in Table 4.3. Then, to calculate the state for each recognized utterance, we take the maximum probable state, using Eq. (4.3). For instance, the user intent for $u_2$ is learned as $t$, i.e., *transports*.

Finally, the transition model can be learned using Eq. (4.4). This is a maximum likelihood transition model. Figure 4.3 shows graphically that we use the maximum likelihood transition model, which is learned based on the learned states (intents), denoted by $s$, and the extracted actions from the dialog set, denoted by $a$.

Note, not every possible triple $(s_1, a_1, s_2)$ does occur in the data, so some of the probabilities in Eq. (4.4) could be zero. We avoid this by adding one to the numerator in Eq. (4.4), a technique known as *smoothing*. In Eq. (4.5) we add 1, as many as count of $(z_1, a_1, z_2)$, in the numerator, so we should add $\text{Count}(z_1, a_1, z_2)$ to the denominator so that it sums to one. Therefore, the transition model can be calculated as:

$$T(s_1, a_1, s_2) = \Pr(z'|z, a) = \frac{\text{Count}(z_1, a_1, z_2) + 1}{\text{Count}(z_1, a_1) + \text{Count}(z_1, a_1, z_2)}. \tag{4.5}$$

Thus, we use Eq. (4.5) for learning the transition model of the dialog POMDP.

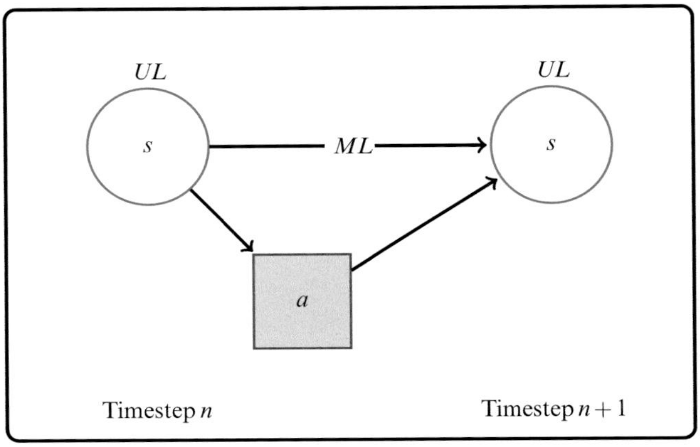


**Fig. 4.3** The maximum likelihood transition model is learned using the extracted actions, $a$, represented using the *shaded square*, and the learned states, $s$, represented in the *light circles*

The transition model introduced in Eq. (4.5) is similar to the user goal model for the factored transition model in Eq. (3.7), proposed by Williams and Young (2007) and Williams (2006). More recently, Lison (2013) proposed a model-based Bayesian reinforcement learning approach to estimate the transition models for dialog management. The method explicitly represents the model uncertainty using a posterior distribution over the model parameters. In contrast to the previous works, we learn such user model from dialogs, as described in Sect. 4.2.1, assign them to the recognized utterances by Eq. (4.3), and then learn the smoothed maximum likelihood user model using Eq. (4.5).

## 4.4 Learning Observations and Observation Model

In this section, we go through the third step in the descriptive Algorithm 1. That is, reducing the observations significantly and learning the observation model. In this context, the definition of observations and observation model can be non-trivial. In particular, the time complexity for learning the optimal policy of a POMDP is double exponential to the number of observations (Cassandra et al. 1995). In non-trivial domains such as ours, the number of observations is large. Depending on the domain, there can be hundreds or thousands of words which ideally should be used as observations. In this case, solving a POMDP with that many observations is intractable.

Therefore, in order to be able to apply POMDPs in such domains, we need to reduce the number of observations significantly. We learn an intent observation model based on HTMM. Figure 4.4 shows that the intent observations, denoted by $o$, are learned based on an unsupervised learning technique and added to the learned models. Before we propose the intent observation model, we introduce the keyword observation model.

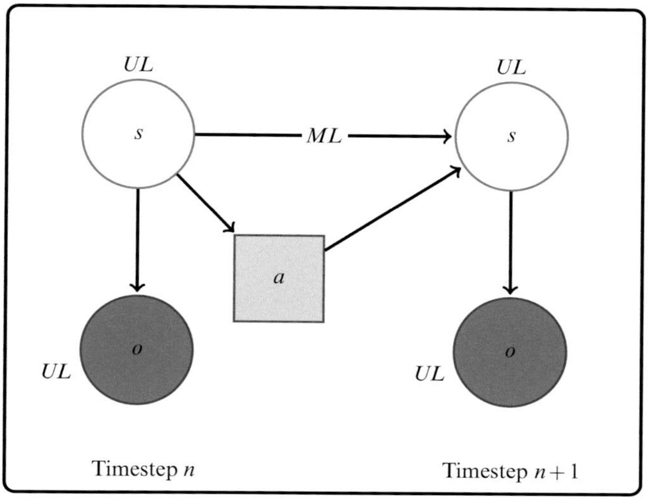


**Fig. 4.4** The observations, $o$, are learned based on an unsupervised learning (UL) method, and are represented using the *shaded circles*

### 4.4.1 Keyword Observation Model

For each state, this model uses the 1-top keyword which best represents the state. For instance, for SACTI-1 dialogs the 1-top keyword in Table 4.2 are the observations which include *hotel, street*, and *restaurant*. These observations can best represent the states: *visits, transports*, and *foods*, respectively. Moreover, an auxiliary observation, which is called *confusedObservation*, is used, when none of the keyword observations occurs in a recognized user utterance. If an utterance includes more than one of the keyword observation, the *confusedObservation* is also used as the observation.

For the keyword observation model, we define a maximum likelihood observation model:

$$\Omega(o', a, s') = \Pr(o'|a, s') = \frac{\text{Count}(a, s', o')}{\text{Count}(a, s')}.$$

To make a more robust observation model, we apply smoothing to the maximum likelihood observation model for instance $\delta$ smoothing where $0 \leq \delta \leq 1$. We set $\delta$ to 1 to have add-1 smoothing:

$$\Omega(o', a, s') = \Pr(o'|a, s') = \frac{\text{Count}(a, s', o') + 1}{\text{Count}(a, s') + \text{Count}(a, s', o')}.$$

In the experiment of the observation models, in Sect. 6.2.2, the dialog POMDP with the keyword observation model is called *keyword POMDP*.

### 4.4.2 Intent Observation Model

Given the recognized user utterance $\tilde{u} = [w_1, \ldots, w_n]$, the observation $o$ is defined in the same way as the state, i.e., the highest probable underlying intent in Eq. (4.3). So the observation $o$ would be:

$$o = \operatorname*{argmax}_{z} \prod_{w_i} \Pr(w_i|z). \tag{4.6}$$

Recall that $\Pr(w_i|z)$ is learned and stored in the vector $\boldsymbol{\beta}_{w_i z}$ from Eq. (4.2).

Notice that for the intent model, each state itself is the observation. As such, the set of observation is equivalent to the set of states. For instance, for SACTI-1 example the intent observations are *vo, to*, and *fo*, respectively, for *visits, transports*, and *foods* states.

Similar to the keyword model, the intent observation model can be defined as:

$$\Omega(o', a, s') = \Pr(o'|a, s') = \frac{\text{Count}(a, s', o')}{\text{Count}(a, s')}.$$

Note that in the intent observation model, we essentially end up with an MDP model. This is because we use the highest probable intent as state and we use the highest probable intent as observation as well. So, we end up with a deterministic observation model, which is such as an MDP as discussed in Sect. 3.1.2. However, we can use a sort of smoothing to allow a small probability for other observations than the observation corresponding to the current state. In the experiment of the observation models, Sect. 6.2.2, we use the intent model without smoothing as the learned *intent MDP* model.

Additionally, we can estimate the intent observation model using the recognized utterances $\tilde{u}$ inside the training dialog $\boldsymbol{d}$, and using the vector $\boldsymbol{\beta}_{wz}$ and $\boldsymbol{\theta}_z$, reflected in Eqs. (4.2) and (4.1), respectively. Assume that we want to estimate $\Pr(o')$ in which $o'$ is drawn from Eq. (4.6), then we have:

$$\begin{aligned}\Pr(o') &= \sum_{w} \Pr(w, o') \\ &= \sum_{w} \Pr(w|o')\Pr(o') \\ &= \sum_{w} \boldsymbol{\beta}_{wo'} \boldsymbol{\theta}_{o'}.\end{aligned} \tag{4.7}$$

To estimate $\Pr(o'|a, s')$, the multiplication in Eq. (4.7) is performed only after visiting the action state pair $(a, s')$. Therefore, we use this calculation to learn the intent observation model. In the experiment of the observation models, Sect. 6.2.2, the dialog POMDP with the intent observation model is called *intent POMDP*.

Atrash and Pineau (2010) proposed a Bayesian method of learning an observation model for POMDPs. Their observation model also draws from a Dirichlet distribution whose parameters are updated when the POMDP action matches with that of expert. More specifically, their proposed algorithm samples a few POMDPs of which only the observation models are different. Then, it learns the policy of each POMDP and go through a few runs by receiving an observation and performing the action of each POMDP. When the action of a POMDP matches with that of expert, observation model of that POMDP is updated. The $n$ worst POMDP models are eliminated and then $n$ new POMDP models are sampled. This process continues until the algorithm is left with a few POMDPs in which the actions match highly with those of experts.

The work presented in Atrash and Pineau (2010) is different from ours as their work is a sample-based Bayesian method. That is, $n$ models are sampled and after updating each model, each POMDP model is solved, and the POMDP models are kept in which actions matched to the expert actions. The proposed observation models in this book, however, learns from expert/machine dialogs; it directly learns the observation model from dialogs and then learns the policy of the learned POMDP model.

As mentioned in Sect. 3.2.2, Png and Pineau (2011) proposed a Bayesian approach for updating the observation model of SmartWheeler dialog POMDP. Similar to Ross et al. (2007, 2011), Png and Pineau (2011) used a Bayes-Adaptive POMDP for learning the observation model. More specifically, they considered a parameter for Dirichlet counts inside the POMDP state model. As such, when the POMDP updates its belief it also updates the Dirichlet counts which subsequently leads to the update of the observation model. As opposed to Png and Pineau (2011), we learned the model totally from SmartWheeler dialogs. Moreover, our idea of observations is based on intents or keywords that is learned from dialogs, whereas observations in Png and Pineau (2011) are given/assumed.

In our previous work (Chinaei et al. 2012), we applied the two observation models on SACTI-1 and SmartWheeler dialogs. Our experimental results showed that the intent observation model outperforms the keyword observation model, significantly, based on accumulated mean rewards in simulation runs. In Chap. 6, we show the two learned models on SmartWheeler dialogs and present the results. In the following section, we go through the illustrative example on SACTI-1, and learn a dialog POMDP by application of the proposed methods of this chapter on SACTI-1 dialogs.

## 4.5 Example on SACTI Dialogs

We use the proposed methods in Sects. 4.2–4.4 to learn a dialog POMDP from SACTI-1 dialogs. First, we use the learned intents in Table 4.2 as states of the domain. Based on the captured intents, we defined three non-terminal states for the SACTI-1 machine as follows:

1. *visits (v)* ,
2. *transports (t)* ,
3. *foods (f)*.

Moreover, we defined two terminal states:

4. *success*,
5. *failure*

The two terminal states are for dialogs which end successfully and unsuccessfully (respectively). The notion of successful or unsuccessful dialog is defined by user. In SACTI-1, the user assigns the level of precision and recall of the received information, after finishing each dialog. This is the only explicit feedback that we require to define the terminal states of the dialog POMDP. A dialog is successful if its precision and recall are above a predefined threshold.

The set of actions comes directly from the SACTI-1 dialog set, and they include:

1. INFORM,
2. REQUEST,
3. GREETINGFAREWELL,
4. REQREPEAT,
5. STATEINTERP,
6. INCOMPLETEUNKNOWN,
7. REQACK,
8. EXPLACK,
9. HOLDFLOOR,
10. UNSOLICITEDAFFIRM,
11. RESPONDAFFIRM,
12. RESPONDNEGATE,
13. REJECTOTHER,
14. DISACK.

For instance, GREETINGFAREWELL is used for initiating or ending a dialog, INFORM is used for giving information for a user intent, REQACK is used for the machine request for user acknowledgement; STATEINTERP is used for interpreting the intents of user. Using such states and actions, the transition model of our dialog POMDP was learned based on the method in Sect. 4.3.

The observations for SACTI-1 would be *hotel, street, restaurant, confusedObservation, success, failure* in the case of keyword observation model, and the observations would be *vo, to, fo, success, failure* in the case of intent observation model. Then, based on the proposed methods in Sect. 4.4, both keyword and intent observation models are learned. As mentioned in the previous section, the intent POMDP with the deterministic observation model is the intent MDP, which is used for the experiments of Chaps. 5 and 6.

For our experiments, we used a typical reward function. Similar to previous work, we penalized each action in non-terminal states by $-1$, i.e., $-1$ reward for each

**Table 4.4** Results of applying the two observation models on the SACTI-1 sample

| | |
|---|---|
| | ... |
| $u_1$ | yeah hello this is johan schmulka uh and i'm uh searching for a bar in this town can you may be tell me where the cafe blu is |
| $\tilde{u}_1$ | [hello this is now seven four bus and do you tell me where to cafe blu is] |
| $o_1$ | *confusedObservation (fo)* |
| $a_1$: | INFORM(*foods*) |
| $m_1$ | cafe blu is on alexander street |
| $u_2$ | oh um yeah how can i get to alexander street and where exactly is it i know there a shopping area on alexander street um |
| $\tilde{u}_2$ | [i am yeah i am at the alexander street and where is it was on a the center of alexander street] |
| $o_2$ | *street (to)* |
| $a_2$: | INFORM(*transports*) |
| $m_2$ | it is on the east side of alexander street so %um it's %um just off middle road |
| | ... |

dialog turn (Williams and Young 2007). Moreover, actions in the *success* terminal state receive +50 as reward and actions in the *failure* terminal state receive −50 as reward.

Table 4.4 represents the sample from SACTI-1, introduced in Table 4.1, after applying the two observation models on the dialogs. The first user utterance is shown in $u_1$. Note that $u_1$ is hidden to the machine and is recognized as the line in $\tilde{u}_1$. Then, $\tilde{u}_1$ is reduced and received as the observation in $o_1$; if the keyword observation model is used, the observation will be *confusedObservation*. This is because none of the keywords *hotel, street*, and *restaurant* occur in $\tilde{u}_1$. But, if the intent observation model is used then the observation inside parenthesis is used, i.e., *fo* which is an observation with high probability for *foods* state, and with small probability for *visits* and *transports* states.

The next line, $a_1$ shows the machine action in the form of dialog acts. For instance, INFORM(*foods*) is the machine dialog act which is uttered by the machine as $m_1$, i.e., *cafe blu is on alexander street*. Next, the table shows $u_2$, $\tilde{u}_2$, $o_2$, and $a_2$. Note that in $o_2$, as opposed to $o_1$ in the case of keyword observation model, the keyword *street* occurs in the recognized utterance $\tilde{u}_2$.

### 4.5.1 HTMM Evaluation

We evaluated HTMM for learning user intents in dialogs. To achieve that, we measured the performance of the model on the SACTI data set based on the definition of *perplexity* similar to Blei et al. (2003) and Gruber et al. (2007). For a learned topic model on a train data set, perplexity can be considered as a measure of on average how many different equally probable words can follow any given word. Therefore, it measures how difficult it is to estimate the words from the model. So, the lower the perplexity is, the better is the model.

Formally, the perplexity of a test dialog $\boldsymbol{d}$ after observing the first $k$ words can be drawn using the following equation:

$$\text{perplexity} = \exp\left(-\frac{\log\ \Pr(w_{k+1}, \ldots, w_{|\boldsymbol{d}|} | w_1, \ldots, w_k)}{|\boldsymbol{d}| - k}\right).$$

We can manipulate the probability distribution in the equation above as:

$$\Pr(w_{k+1}, \ldots, w_{|\boldsymbol{d}|} | w_1, \ldots, w_k) = \sum_i^N \Pr(w_{k+1}, \ldots, w_{|\boldsymbol{d}|} | z_i) \Pr(z_i | w_1, \ldots, w_k),$$

where $z_i$ is a user intent in the set of $N$ captured user intents from the train set. Given a user intent $z_i$, the probability of observing $w_{k+1}, \ldots, w_{|\boldsymbol{d}|}$ are independent of each other, so we have:

$$\Pr(w_{k+1}, \ldots, w_{|\boldsymbol{d}|} | w_1, \ldots, w_k) = \sum_i^N \prod_{j=k+1}^{|\boldsymbol{d}|} \Pr(w_j | z_i) \Pr(z_i | w_1, \ldots, w_k).$$

To find out the perplexity, we learned the intents for each test dialog $\boldsymbol{d}$ based on the first $k$ observed words in $\boldsymbol{d}$, i.e., $\boldsymbol{\theta}_{\text{new}} = \Pr(z_i | w_1, \ldots, w_k)$ is calculated for each test dialog, whereas the vector $\boldsymbol{\beta}$, which retains $\Pr(w_j | z_i)$ [cf. Eq. (4.2)], is learned from the training dialogs. We calculated the perplexity for 5 % of the dialogs in data set and we used the 95 % rest for training. Figure 4.5 shows the average perplexity after observing the first $k$ utterances of test dialogs. As the figure shows, the perplexity is reduced significantly when we observe new utterances.

At the end of Sect. 4.2.1 we mentioned that HTMM has a small computation time since it has a special form of the transition matrix (Gruber et al. 2007; Gruber and Popat 2007). Here we show the convergence rate of HTMM based on the convergence of log likelihood of data. Figure 4.6 shows the log likelihood of the observations for 30 iterations of the algorithm. We can see in the figure that the algorithm converges quite fast. For the given observations, the log likelihood is computed by averaging over possible intents:

$$\hat{\boldsymbol{\theta}}_{\text{MLE}} = \sum_{i=1}^{|\boldsymbol{D}|} \sum_{j=1}^{|\boldsymbol{d}_i|} \log \sum_{t=1}^{N} \Pr(w_{i,j} = w | z_{i,j} = z_t).$$

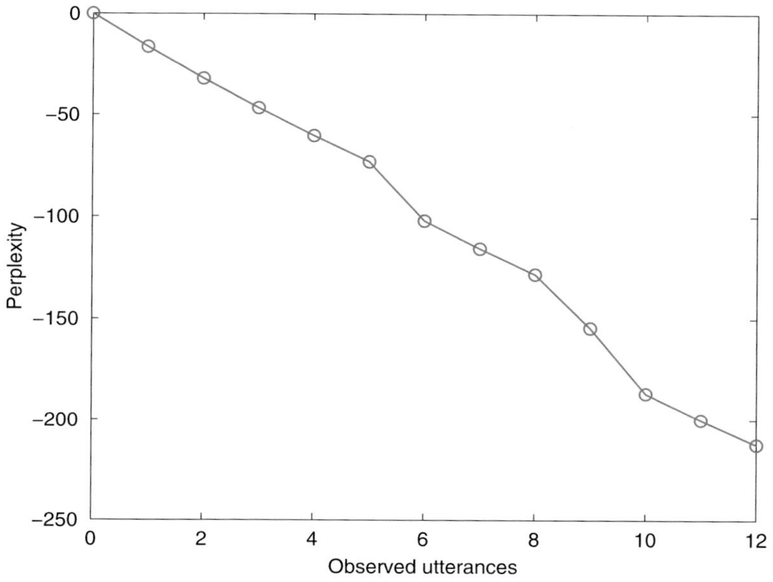


**Fig. 4.5** Perplexity trend with respect to increase of the number of observed user utterances

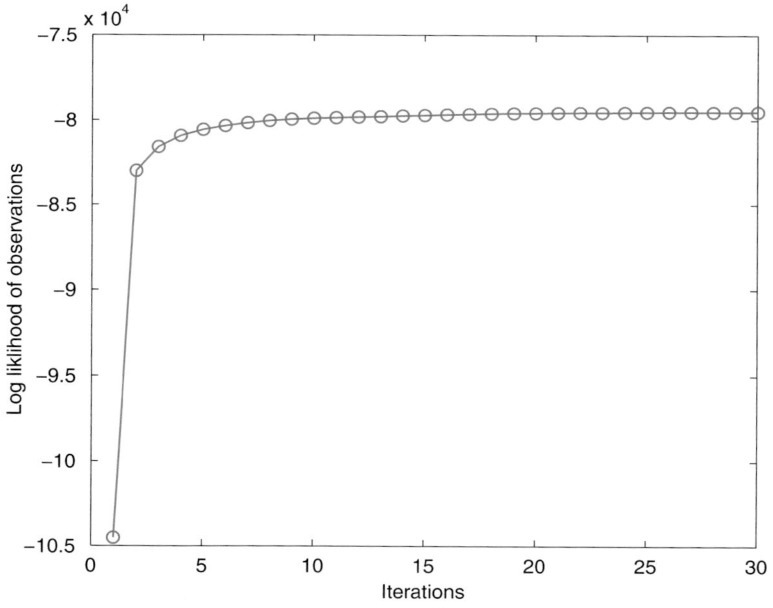


**Fig. 4.6** Log likelihood of observations in HTMM as a function of the number of iterations

### 4.5.2 Learned POMDP Evaluation

We evaluated the learned intent POMDP from SACTI-1 dialogs, introduced in Sect. 4.2.2, using simulation runs. These results have been presented in our previous work (Chinaei and Chaib-draa 2011). The learned intent dialog POMDP models from SACTI-1 consist of three non-terminal states and two terminal states, fourteen actions, and five intent observations. We solved our POMDP models, using the ZMDP software available online at: http://www.cs.cmu.edu/~trey/zmdp/. We set a uniform distribution on the three non-terminal states, *visits*, *transports*, and *foods*, and set the discount factor to 0.90.

Based on simulation runs, we evaluated the robustness of the learned POMDP models to the ASR noise. There are four levels of ASR noise in SACTI data: *none*, *low*, *medium*, and *high* noise. For each noise level, we randomly took 24 available expert dialogs, calculated the average accumulated rewards for the experts from the 24 expert dialogs, and made a dialog POMDP model from the 24 expert dialogs. Then, for each learned POMDP we performed 24 simulations and calculated their average accumulated rewards. In our experiments, we used the default simulation in the ZMDP software.

Figure 4.7 plots the average accumulated rewards as the noise level changes from 0 to 3 for none, low, medium, and high levels of noise (respectively). As the figure shows, the dialog POMDP models are robust to the ASR noise levels. That is,

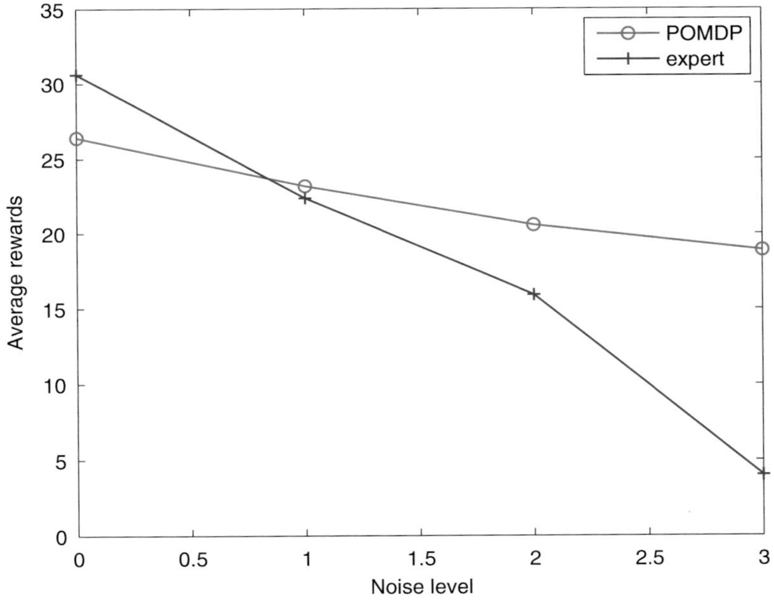


**Fig. 4.7** Average rewards accumulated by the learned dialog POMDPs with respect to the noise level

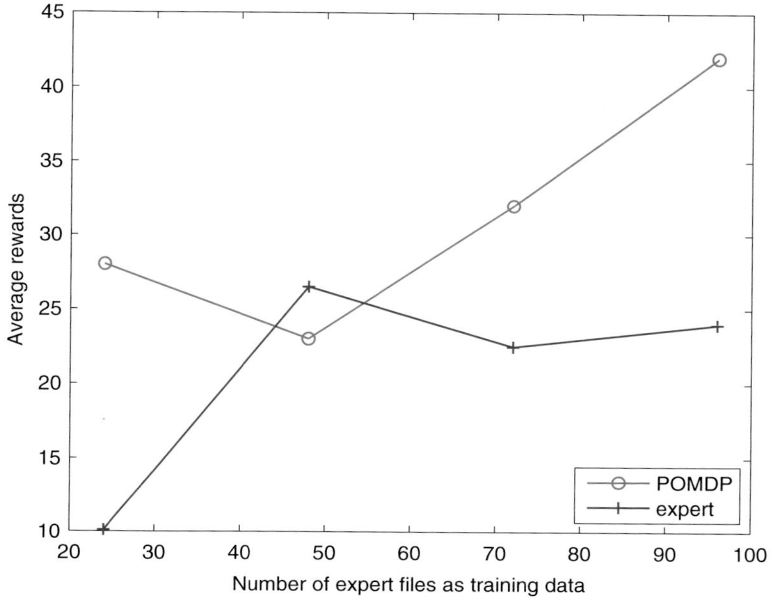


**Fig. 4.8** Average rewards accumulated by the learned dialog POMDPs with respect to the size of expert dialogs as training data

performance of the learned dialog POMDPs decreases only slightly as the noise level increases. On the other hand, performance of experts decreases significantly, in particular at high level of noise. Note in Fig. 4.7 that average accumulated mean reward for the experts is highest when there is no noise, and it is higher than the subsequent learned POMDPs. This is reasonable as the human expert can have best performance in the least uncertain conditions, i.e., when there is no noise.

Moreover, we evaluated the performance of the learned dialog POMDPs as a function of expert dialogs (as training data), shown in Fig. 4.8. Similar to the previous experiments, we calculated the average accumulated rewards for the learned POMDPs and for the experts from the subsequent expert dialogs. Overall, performance of the learned dialog POMDPs is directly related to the number of expert dialogs and we find that more training data implies better performance.

Table 4.5 shows a sample from the learned dialog POMDP simulation. The first action, $a_1$, is generated by dialog POMDP, which is shown in the form of natural language in the following line, denoted by $m_1$. Then, the observation $o_2$ is generated by environment, *vo*. For instance, the recognized user utterance could have been an utterance such as: $\tilde{u}$ : *I would like a hour there museum first*, and therefore its intent observation can be calculated using Eq. (4.6). Notice that these results are only based on the dialog POMDP simulation, where there exists neither user utterance nor machine's utterance but only the simulated action and observations. Then, based on

**Table 4.5** A sample from SACTI-1 dialog POMDP simulation

| | |
|---|---|
| $a_1$: | GREETINGFAREWELL |
| $m_1$: | How can I help you? |
| $o_2$: | *vo* |
| $b_1$: | *t:0.04814 v:0.91276 f:0.03909* |
| | |
| $a_2$: | INFORM(*visits*) |
| $m_2$: | Here is information about visiting areas |
| $o_2$: | *to* |
| $b_2$: | *t:0.96732 v:0.00818 f:0.02449* |
| | |
| $a_3$: | INFORM(*transports*) |
| $m_3$: | *Here is information about transportation* |
| $o_3$: | *to* |
| $b_3$: | *t:0.99385 v:0.00031 f:0.00583* |
| | |
| $a_4$: | REQACK(*transports*) |
| $m_4$: | *Are you looking for transportation* |
| $o_4$: | *to* |
| $b_4$: | *t:0.94565 v:0.04833 f:0.00600* |
| | |
| $a_5$: | INFORM(*transports*) |

the received observation the POMDP belief, shown in $b_1$, is updated, using Eq. (3.3). Based on belief $b_1$, the dialog POMDP performs the next action, denoted by $a_2$.

In turns 3 to 5 shown in Table 4.5, we can see that the learned dialog POMDP performs intuitively. In turn 3, the dialog POMDP informs the user about *transports*, after receiving the observation *to* in turn 2 (the observation for *transports*). In $a_4$, the dialog POMDP requests for acknowledgement that the user actually looks for *transports*, perhaps since it has already informed the user about *transports* in turn 3. After receiving the observation *to* in turn 4, and updating the belief, the dialog POMDP informs the user again about *transports* in $a_5$.

## 4.6 Conclusions

In this chapter, we introduced methods for learning the dialog POMDP states, transition model, observations and observation model, from recognized user utterances. In the intent-based dialog domains in which the user intent is the dialog state, an interesting problem is to learn the user intents from unannotated user utterances.

To do so, first we studied HTMM, an unsupervised topic modeling approach that adds Markovian property to the LDA model. We then applied the HTMM method on dialogs to learn the set of user intents and thus the probability distribution of user intents for each recognized user utterance. We then made use of the learned user intents as the dialog POMDP states and learned a smoothed maximum likelihood transition model. Furthermore, we proposed two sets of observations: keyword and intent observations, automatically learned from dialogs, as well as their subsequent observation models.

Throughout this chapter, we applied the proposed methods on SACTI dialogs; we then evaluated the HTMM method for learning user intents using SACTI dialogs, based on the definition of perplexity. Finally, we evaluated the learned intent dialog POMDPs in simulation runs based on average accumulated rewards. The simulation results show that the learned intent dialog POMDPs are robust to the ASR noise.

Building on the learned dialog POMDP model components, in the next chapter, we propose two algorithms for learning the reward function based on IRL techniques.

# Chapter 5
# Learning the Reward Function

## 5.1 Introduction

In Sect. 3.1, we introduced reinforcement learning (RL) as a technique for learning policy in stochastic/uncertain domains. In this context, RL works by optimizing a *defined* reward function in the (PO)MDP framework. In particular, choice of the reward function has been usually hand-crafted based on the domain expert intuition. However, it is evidently more convenient for the expert to demonstrate the policy. Thus, recently the inverse reinforcement learning (IRL) method is used to approximate the reward function that some expert agent appears to be optimizing.

Recall Fig. 3.1 which showed the interaction between a machine and its environment. We present again the figure here, this time with more details in Fig. 5.1. In this figure, circles represent learned models. The model denoted by POMDP includes the POMDP model components (without a reward function) which have been learned from introduced methods in Chap. 4. The learned POMDP together with action/observation *trajectories* are used in IRL to learn the reward function, denoted by R. Then, the learned POMDP and reward function are used in a POMDP solver to learn/update the optimal policy.

In this chapter, we introduce IRL and propose POMDP-IRL algorithms for the fourth step of the descriptive Algorithm 1: learning the reward function based on IRL techniques and using the learned POMDP model components (Chinaei and Chaib-Draa 2014a). In this context, Ng and Russell (2000) proposed multiple IRL algorithms in the MDP framework that work by maximizing the sum of the margin between the policy of the expert (agent) and the intermediate *candidate* policies. These algorithms account for the case in which the expert policy is represented *explicitly* and the case where the expert policy is known only through observed expert *trajectories*.

IRL in POMDPs, in short POMDP-IRL, is particularly challenging due to the difficulty in solving POMDPs as discussed in Sect. 3.1.2. Recently, Choi and Kim (2011) proposed POMDP-IRL algorithms by extending MDP-IRL algorithms of



H. Chinaei, B. Chaib-draa, *Building Dialogue POMDPs from Expert Dialogues*, SpringerBriefs in Speech technology, DOI 10.1007/978-3-319-26200-0_5

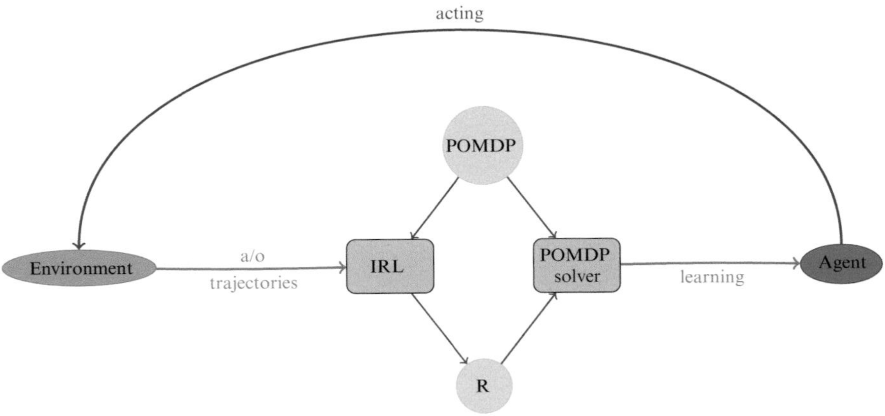


**Fig. 5.1** The cycle of acting/learning between the agent and environment. The *circles* represent the models. The model denoted by POMDP includes the POMDP model components, without a reward function, learned from introduced methods in Chap. 4. The learned POMDP model together with action/observation trajectories are used in IRL to learn the reward function denoted by R. The learned POMDP and reward function are used in the POMDP solver to learn/update the policy

Ng and Russell (2000) to POMDPs. In particular, Choi and Kim (2011) provided a general framework for POMDP-IRL by modeling the expert policy as a finite state controller (FSC) and thus using point-based policy iteration (PBPI) (Ji et al. 2007) as POMDP solver. The trajectory-based algorithms in Choi and Kim (2011) also required the FSC-based POMDP solvers (PBPI). In particular, they proposed a trajectory-based algorithm called max-margin between values (MMV) for the POMDP framework. Since such algorithms spent most of the time solving the intermediate policies, they suggested modifying the trajectory-based algorithms to be able to use other POMDP solvers such as Perseus (Spaan and Vlassis 2005), etc.

In this chapter, we extend the trajectory-based MDP-IRL algorithm of Ng and Russell (2000) to POMDPs. We assume that the model components are known, similar to Ng and Russell (2000) and Choi and Kim (2011). Fortunately, in dialog management, the transition and observation models can be calculated from Wizard-of-Oz data (Choi and Kim 2011) or a real system data, as mentioned in Sect. 1.1. In particular, in Chap. 4, we proposed methods for learning such components from data and showed the illustrative example of learning the dialog POMDP model components from SACTI-1 dialogs, collected in a Wizard-of-Oz setting (Williams and Young 2005). Then, the learned dialog POMDP model together with expert dialog trajectories can be used in IRL algorithms to learn a reward function for the expert policy.

In this context, IRL is an *ill-posed* problem. That is, there is not a single reward function that makes expert policy optimal, but infinitely many of them. We show this graphically in Chinaei (2010), through an experiment on a toy dialog MDP. Since there are many reward functions that make the expert policy optimal, one approach is based on linear programming to find one of the possible solutions. The linear program constrains the set of possible reward functions where the rewards are represented as a linear representation of dialog features, and finds a solution among the limited set of solutions.

Note that in (PO)MDP-IRL the expert is assumed to be a (PO)MDP expert. That is, the expert policy is the policy that the underlying (PO)MDP framework optimizes. Similar to the previous work, we perform our IRL algorithm on (PO)MDP experts in this book.

In Sect. 5.2, we introduce the basic definitions of IRL. In this section, we also study in detail the main trajectory-based IRL algorithm for MDPs, introduced by Ng and Russell (2000). We call this algorithm MDP-IRL. The material in Sect. 5.2 makes the foundation on which Sect. 5.3 is built. In particular, in Sect. 5.3.1 we propose a trajectory-based IRL algorithm for POMDPs, called POMDP-IRL-BT, which is an extension of the MDP-IRL algorithm of Ng and Russell (2000) for POMDPs. Then, in Sect. 5.3.2 we describe a point-based IRL algorithm for the POMDP framework, called PB-POMDP-IRL. In Sect. 5.4, we go through IRL related work, particularly for POMDPs. In Sect. 5.6, we revisit the SACTI-1 example; we apply the POMDP-IRL-BT and PB-POMDP-IRL algorithms on the learned dialog POMDP from SACTI-1 (introduced in Sect. 4.5) and compare the results. Finally, we conclude this chapter in Sect. 5.7.

## 5.2 IRL in the MDP Framework

In IRL, given an expert policy and an underlying MDP, the problem is to learn a reward function that makes the expert policy optimal. That is, given the expert policy, approximate a reward function for the MDP such that the optimal policy of the MDP includes the expert policy. In this section, we describe IRL for MDPs (MDP-IRL) using expert trajectories, represented as $(s_0, \pi_E(s_0), \ldots, s_{|S|-1}, \pi_E(s_{|S|-1}))$. To begin let us introduce the following definitions:

- an *expert reward function*, denoted by $R^{\pi_E}$, is an unknown reward function for which the optimal policy is *expert policy*. We have the following definitions:
  - the *expert policy*, denoted by $\pi_E$, is a policy of the underlying MDP that optimizes the expert reward function $R^{\pi_E}$,
  - the *value of the expert policy*, denoted by $V^{\pi_E}$, is the value of the underlying MDP in which the reward function is the expert reward function $R^{\pi_E}$.
- a *candidate reward function*, denoted by $R$, is a reward function that could potentially be the expert reward function. We have the following definitions:
  - the *candidate policy*, denoted by $\pi$, is a policy of the underlying MDP that optimizes the candidate reward function $R$,
  - the *value of the candidate policy*, denoted by $V^{\pi}$, is the value of the candidate policy $\pi$ that optimizes the candidate reward $R$.

Then, IRL aims to find a reward function in which the expert's policy is both optimal and maximally separated from other policies. To do this, some candidate reward functions and their subsequent policies are generated from the expert's behavior. The candidate reward function is approximated by maximizing the value of the expert policy with respect to all previous candidate policies. The new candidate reward function and policy are then used to approximate another new set of models. This process iterates until the difference in values of successive candidate policies is less than some threshold. The final candidate reward function is the solution to the IRL task.

Formally, we formulate the IRL problem as an MDP without a reward function, denoted by $\text{MDP}\backslash R = \{S, A, T_a, \gamma\}$, so that we can calculate the optimal policy of the MDP given any choice of candidate reward function. Having $t$ candidate policies $\pi_1, \ldots, \pi_t$, the next candidate reward is estimated by maximizing $\boldsymbol{d}^t$, the sum of the margins between value of expert policy and each learned candidate policy. Then, the objective function is as follows:

$$\text{maximize}\, \boldsymbol{d}^t = (\boldsymbol{v}^{\pi_E} - \boldsymbol{v}^{\pi_1}) + \cdots + (\boldsymbol{v}^{\pi_E} - \boldsymbol{v}^{\pi_t}), \tag{5.1}$$

where $\boldsymbol{v}^{\pi}$ is the vector representation for value function:

$$\boldsymbol{v}^{\pi} = (\boldsymbol{v}^{\pi}(s_0), \ldots, \boldsymbol{v}^{\pi}(s_{|S|-1}))$$

and $\boldsymbol{v}^{\pi}(s_i)$ is the value of state $s_i$ under policy $\pi$, which can be drawn from Eq. (3.1). That is, we have:

$$\boldsymbol{v}^{\pi} = \boldsymbol{r}^{\pi} + \gamma \boldsymbol{T}^{\pi} \boldsymbol{v}^{\pi}, \tag{5.2}$$

where

- $\boldsymbol{v}^{\pi}$ is a vector of size $|S|$ in which $\boldsymbol{v}^{\pi}(s) = V^{\pi}(s)$.
- $\boldsymbol{r}^{\pi}$ is a vector of size $|S|$ in which $\boldsymbol{r}^{\pi}(s) = R(s, \pi(s))$.
- $\boldsymbol{T}^{\pi}$ is the transition matrix for policy $\pi$, that is a matrix of size $|S| \times |S|$ in which $\boldsymbol{T}^{\pi}(s, s') = T(s, \pi(s), s')$.

Notice that in IRL it is assumed that the reward of any state $s$ can be represented as the linear combination of some features of state $s$, such as a feature vector defined as:

$$\boldsymbol{\phi} = (\phi_1(s, a), \ldots, \phi_K(s, a)),$$

where $K$ is the number of features and each feature $\phi_i(s, a)$ is a basis function for the reward function. The reward function can be shown as the multiplication of two vectors $\boldsymbol{\Phi}^\pi$ and $\boldsymbol{\alpha}$ as:

$$\boldsymbol{r}^\pi = \boldsymbol{\Phi}^\pi \boldsymbol{\alpha}, \tag{5.3}$$

where $\boldsymbol{\alpha} = (\alpha_1, \ldots, \alpha_K)$ are feature weights, and $\boldsymbol{\Phi}^\pi$ is a matrix of size $|S| \times K$ consisting of state action features for policy $\pi$, defined as:

$$\boldsymbol{\Phi}^\pi = \begin{pmatrix} \boldsymbol{\phi}(s_0, \pi(s_0))^{\mathrm{T}} \\ \ldots \\ \boldsymbol{\phi}(s_{|S|-1}, \pi(s_{|S|-1}))^{\mathrm{T}} \end{pmatrix}.$$

For the expert policy $\pi_E$, the state action features become:

$$\boldsymbol{\Phi}^{\pi_E} = \begin{pmatrix} \boldsymbol{\phi}(s_0, \pi_E(s_0))^{\mathrm{T}} \\ \ldots \\ \boldsymbol{\phi}(s_{|S|-1}, \pi_E(s_{|S|-1}))^{\mathrm{T}} \end{pmatrix}.$$

We can manipulate Eq. (5.2):

$$\begin{aligned} \boldsymbol{v}^\pi &= \boldsymbol{r}^\pi + \gamma \boldsymbol{T}^\pi \boldsymbol{v}^\pi \\ \boldsymbol{v}^\pi - \gamma \boldsymbol{T}^\pi \boldsymbol{v}^\pi &= \boldsymbol{r}^\pi \\ (\boldsymbol{I} - \gamma \boldsymbol{T}^\pi)\boldsymbol{v}^\pi &= \boldsymbol{r}^\pi \\ \boldsymbol{v}^\pi &= (\boldsymbol{I} - \gamma \boldsymbol{T}^\pi)^{-1} \boldsymbol{r}^\pi. \end{aligned}$$

Therefore, from the last equality we have:

$$\boldsymbol{v}^\pi = (\boldsymbol{I} - \gamma \boldsymbol{T}^\pi)^{-1} \boldsymbol{r}^\pi. \tag{5.4}$$

Using Eq. (5.3) in Eq. (5.4), we have:

$$\begin{aligned} \boldsymbol{v}^\pi &= (\boldsymbol{I} - \gamma \boldsymbol{T}^\pi)^{-1} \boldsymbol{\Phi}^\pi \boldsymbol{\alpha} \\ \boldsymbol{v}^\pi &= \boldsymbol{x}^\pi \boldsymbol{\alpha}, \end{aligned} \tag{5.5}$$

where $\boldsymbol{x}^\pi$ is a matrix of size $|S| \times K$ defined as:

$$\boldsymbol{x}^\pi = (\boldsymbol{I} - \gamma \boldsymbol{T}^\pi)^{-1} \boldsymbol{\Phi}^\pi. \tag{5.6}$$

Equation (5.5) shows that the vector of values $\boldsymbol{v}^\pi$ can be represented as multiplication of the feature weight vector $\boldsymbol{\alpha}$ and another vector $\boldsymbol{x}^\pi$.

Similar to Eq. (5.5), for the expert policy $\pi_E$, we have:

$$\boldsymbol{v}^{\pi_E} = \boldsymbol{x}^{\pi_E}\boldsymbol{\alpha}, \tag{5.7}$$

where $\boldsymbol{x}^{\pi_E}$ is a matrix of size $|S| \times K$ defined as:

$$\boldsymbol{x}^{\pi_E} = (\boldsymbol{I} - \gamma \boldsymbol{T}^{\pi_E})^{-1}\boldsymbol{\Phi}^{\pi_E} \tag{5.8}$$

and $\boldsymbol{T}^{\pi_E}$ is a $|S| \times |S|$ matrix where element $\boldsymbol{T}^{\pi_E}(s_i, s_j)$ is the probability of transiting from $s_i$ to $s_j$ with expert action $\pi_E(s_i)$.

Therefore, both a candidate reward function and its subsequent candidate policy can be represented as multiplication of some feature function and the feature weights $\boldsymbol{\alpha}$ [see Eqs. (5.3) and (5.5)]. This enables us to solve Eq. (5.1) as a linear program. Using Eqs. (5.5) and (5.7) in Eq. (5.1), we have:

$$\begin{aligned} &\text{maximize}_{\boldsymbol{\alpha}} \quad \left[ ((\boldsymbol{x}^{\pi_E} - \boldsymbol{x}^{\pi_1}) + \cdots + (\boldsymbol{x}^{\pi_E} - \boldsymbol{x}^{\pi_t}))\, \boldsymbol{\alpha} \right] \\ &\text{subject to} \quad -1 \leq \alpha_i \leq +1 \; \forall i, 1 \leq i \leq K. \end{aligned} \tag{5.9}$$

Having $t$ candidate policies $\pi_1, \ldots, \pi_t$, IRL estimates the next candidate reward by solving the above linear program. That is, IRL learns a new $\boldsymbol{\alpha}$ which represents a new candidate reward function, $\boldsymbol{r} = \boldsymbol{\Phi}^{\pi_E}\boldsymbol{\alpha}$. This new candidate reward has an "optimal policy" which is the new candidate policy $\pi$.

Algorithm 7 shows the MDP-IRL algorithm introduced in Ng and Russell (2000). This algorithm tries to find the expert reward function given an underlying MDP framework. The idea of this algorithm is that the value of expert policy is required to be higher than the value of any other policy under the same MDP framework. This is the maximization in Line 7 of the algorithm where $\boldsymbol{v}^{\pi_E} = \boldsymbol{x}^{\pi_E}\boldsymbol{\alpha}$ and $\boldsymbol{v}^{\pi_l} = \boldsymbol{x}^{\pi_l}\boldsymbol{\alpha}$ are the value of expert policy and the value of candidate policy $\pi_l$, respectively. Notice that this algorithm maximizes the *sum* of the margins between the value of expert policy $\pi_E$ and the value of other candidate policies $\pi_l$.

Let's go through Algorithm 7 in detail. The algorithm starts by randomly initiating values for $\boldsymbol{\alpha}$ to generate the initial candidate reward function $R^1$ in Line 1. Then, using dynamic programming for the MDP with the candidate reward function $R^1$, the algorithm finds policy of $R^1$, denoted by $\pi_1$. In Line 2, $\pi_1$ is used to construct $\boldsymbol{T}^{\pi_1}$ which is used to calculate $\boldsymbol{x}^{\pi_1}$ from Eq. (5.6). Then, in Line 3, expert policy $\pi_E$ is used to construct $\boldsymbol{T}^{\pi_E}$ which is used to calculate $\boldsymbol{x}^{\pi_E}$ from Eq. (5.8).

**Algorithm 7:** MDP-IRL: inverse reinforcement learning in the MDP framework, adapted from Ng and Russell (2000)

**Input**: $MDP \backslash R = \{S, A, T, \gamma\}$, expert trajectories in the form of $D = \{(s_n, \pi_E(s_n), s'_n)\}$, a vector of features $\boldsymbol{\phi} = (\phi_1, \ldots, \phi_K)$, convergence rate $\epsilon$, and maximum iteration $maxT$

**Output**: Finds reward function $R$ where $R = \sum_i \alpha_i \phi_i(s, a)$ by approximating $\boldsymbol{\alpha} = (\alpha_1, \ldots, \alpha_K)$

1 Choose the initial reward $R^1$ by randomly initializing feature weights $\boldsymbol{\alpha}$;
2 Set $\Pi = \{\pi_1\}$ by finding $\pi_1$ using MDP with candidate reward function $R^1$ and value iteration;
3 Set $X = \{\boldsymbol{x}^{\pi_1}\}$ by calculating $\boldsymbol{x}^{\pi_1}$ using $\boldsymbol{T}^{\pi_1}$ and Equation (5.6);
4 Calculate $\boldsymbol{x}^{\pi_E}$ using $\boldsymbol{T}^{\pi_E}$ and Equation (5.8);
5 **for** $t \leftarrow 1$ *to* $maxT$ **do**
6     Find values for $\boldsymbol{\alpha}$ by solving the linear program:
7         maximize$_{\boldsymbol{\alpha}}$ $\boldsymbol{d}^t = \left[ ((\boldsymbol{x}^{\pi_E} - \boldsymbol{x}^{\pi_1}) + \ldots + (\boldsymbol{x}^{\pi_E} - \boldsymbol{x}^{\pi_t}))\boldsymbol{\alpha} \right]$;
8         subject to $|\alpha_i| \leq 1 \ \forall i \ 1 \leq i \leq K$;
9     $R^{t+1} = \sum_i \alpha_i^t \phi_i(s, a)$;
10     **if** $\max_i |\boldsymbol{\alpha}_i^t - \boldsymbol{\alpha}_i^{t-1}| \leq \epsilon$ **then**
11         return $R^{t+1}$;
12     **end**
13     **else**
14         $\Pi = \Pi \cup \{\pi_{t+1}\}$ by finding $\pi_{t+1}$ using MDP with candidate reward function $R^{t+1}$ and value iteration;
15         Set $X = X \cup \{\boldsymbol{x}^{\pi_{t+1}}\}$ by calculating $\boldsymbol{x}^{\pi_{t+1}}$ using $\boldsymbol{T}^{\pi_{t+1}}$ and Equation (5.6);
16     **end**
17 **end**

From Line 5 to Line 17, MDP-IRL goes through the iterations to learn expert reward function by solving the linear program in Line 7 with the constraints in Line 8. For instance, in the first iteration of MDP-IRL, using the linear programming above, the algorithm finds $\boldsymbol{\alpha}$ which maximizes Eq. (5.9). In Line 9, the learned vector values, $\boldsymbol{\alpha}$, make a candidate reward function $R^2$ which introduces a candidate policy $\pi_2$ in Line 14. Then, in Line 15, $\boldsymbol{T}^{\pi_2}$ is constructed for finding $\boldsymbol{x}^{\pi_2}$ from Eq. (5.6). The algorithm returns to Line 5 to repeat the process of learning a new candidate reward until convergence. In this optimization, we also constrain the value of the expert's policy to be greater than that of other policies in order to ensure that the expert's policy is optimal.

Note that in Ng and Russell (2000) there is a slight different algorithm for trajectory based IRL algorithm. The objective function for learning the reward function of expert maximizes sum of the margin between value of expert policy and that of other policies using a monotonic function $f$. That is, the objective function in Ng and Russell (2000) is as follows:

$$\text{maximize}_{\boldsymbol{\alpha}}\ \boldsymbol{d}^t = \left[ f(\boldsymbol{v}^{\pi_E} - \boldsymbol{v}^{\pi_1}) + \cdots + f(\boldsymbol{v}^{\pi_E} - \boldsymbol{v}^{\pi_t}) \right] \tag{5.10}$$

$$\text{subject to} \qquad |\alpha_i| \leq 1\ \forall i\ 1 \leq i \leq K,$$

where Ng and Russell (2000) set $f(x) = x$ if $f(x) > 0$, otherwise, $f(x) = 2x$ to penalize the cases in which the value of expert policy is less than the candidate policy. The authors selected 2 in $f(x) = 2x$ since it had the least sensitivity in their experiments. The maximization in Eq. (5.9) is similar to the one in Eq. (5.10), particularly when $f(x) = x$ for all $x$.

Moreover, in Ng and Russell (2000) it is suggested to approximate the policy values using *Monte Carlo* estimator. Recall the definition of value function in MDPs, shown in Eq. (3.1), defined as:

$$\begin{aligned} V^{\pi}(s) &= E_{s_t \sim T}\left[ \gamma^0 R(s_0, \pi(s_0)) + \gamma^1 R(s_1, \pi(s_1) + \ldots) | \pi, s_0 = s \right] \\ &= E_{s_t \sim T}\left[ \sum_{t=0}^{\infty} \gamma^t R(s_t, \pi(s_t)) | \pi, s_0 = s \right]. \end{aligned}$$

Using $M$ expert trajectory of size $H$, the value function in MDPs can be approximated using Monte Carlo estimator:

$$\begin{aligned} \hat{V}^{\pi}(s_0) &= 1/M \sum_{m=1}^{M} \sum_{t=0}^{H-1} \gamma^t R(s, a) \\ &= 1/M \sum_{m=1}^{M} \sum_{t=0}^{H-1} \gamma^t \boldsymbol{\alpha}^{\mathrm{T}} \boldsymbol{\phi}(s, a). \end{aligned}$$

The trajectory-based MDP-IRL algorithm in Ng and Russell (2000) has been extended to a model-free trajectory-based MDP-IRL algorithm, called LSPI-IRL, during the authors' collaboration with AT&T research labs in 2011. In the LSPI-IRL algorithm, the candidate policies are estimated using the LSPI (least square policy iteration) algorithm (Lagoudakis and Parr 2003). The LSPI-IRL algorithm is presented in Chinaei (2010).

We then extended the trajectory-based MDP-IRL algorithm of Ng and Russell (2000) to a trajectory-based POMDP-IRL algorithm, called POMDP-IRL-BT, which is presented in Sect. 5.3.1.

## 5.3 IRL in the POMDP Framework

In this section, we propose two IRL algorithms from expert trajectories in the POMDP framework. First in Sect. 5.3.1, we extend the MDP-IRL algorithm of Ng and Russell (2000) to POMDPs by approximating the value of expert policy and that of candidate policies [respectively Eqs. (5.7) and (5.5)] for POMDPs. This is done by fixing the number of beliefs to the expert beliefs available in expert trajectories, and by approximating the expert belief transitions, i.e., the probability of transiting from one expert belief to another after performing an action. The algorithm is called POMDP-IRL-BT (BT for belief transitions). Then, in Sect. 5.3.2, we propose a point-based POMDP-IRL algorithm, called PB-POMDP-IRL.

### *5.3.1 POMDP-IRL-BT*

We extend the trajectory-based MDP-IRL algorithm introduced in previous section to POMDPs. Our proposed algorithm, called POMDP-IRL-BT, considers the situation when expert trajectories are in form of $(a_1, o_1, \dots, a_B, o_B)$, where $B$ is the number of generated expert beliefs. Note that by application of the state estimator function in Eq. (3.3), and an assumed belief $b_0$, say the uniform belief, we can calculate expert beliefs $(b_0, \dots, b_{B-1})$. Thus, expert trajectories can be represented as $(b_0, \pi_E(b_0), \dots, b_{B-1}, \pi_E(b_{B-1}))$.

The POMDP-IRL-BT algorithm is similar to the MDP-IRL algorithm, described in Sect. 5.2, but instead of states we use the finite number of expert beliefs that occurred in expert trajectories. Moreover, we approximate a belief transition for expert beliefs in the place of the transition model in MDPs. More specifically, we approximate the value of the expert policy and the value of candidate policies by approximating Eqs. (5.7) and (5.5), respectively, for POMDPs. Therefore, in IRL for POMDPs we maximize the margin:

$$\boldsymbol{d}^t = (\boldsymbol{v}_b^{\pi_E} - \boldsymbol{v}_b^{\pi_1}) + \cdots + (\boldsymbol{v}_b^{\pi_E} - \boldsymbol{v}_b^{\pi_t}),$$

where $\boldsymbol{v}_b^{\pi_E}$ is an approximation of the value of the expert policy. This expert policy is based on the expert beliefs that occurred in expert trajectories. Moreover, each $v_b^{\pi_t}$ is an approximation of value of the candidate policy $\pi_t$ which is calculated by approximating expert belief transitions.

To illustrate these approximations, consider the value function for POMDPs shown in Eq. (3.5). Using the vector representation, we can rewrite Eq. (3.5) as:

$$\boldsymbol{v}_b^{\pi} = \boldsymbol{r}_b^{\pi} + \gamma \boldsymbol{P}^{\pi} \boldsymbol{v}_b^{\pi}, \tag{5.11}$$

where

- $\boldsymbol{v}_b^\pi$ is a vector of size $B$: the number of expert beliefs in which $\boldsymbol{v}_b^\pi(b) = V^\pi(b)$ (from Eq. (3.5)).
- $\boldsymbol{r}_b^\pi$ is a vector of size $B$ in which $\boldsymbol{r}_b^\pi(b) = R(b, \pi(b))$, where $R(b, a)$ comes from Eq. (3.4).
- $\boldsymbol{P}^\pi$ is a matrix of size $B \times B$ that is the *belief transition* matrix for policy $\pi$, in which:

$$\boldsymbol{P}^\pi(b, b') = \sum_{o' \in O} \Big[ \Pr(o'|b, \pi(b))\ \mathit{ifClosest}((\text{SE}(b, \pi(b), o'), b') \Big], \quad (5.12)$$

where SE is the state estimator function in Eq. (3.3) and $\mathit{ifClosest}(b'', b')$ determines if $b'$ is the closest expert belief to $b''$, the belief created as a result of state estimator function. Formally, we define $\mathit{ifClosest}(b'', b')$ as:

$$\mathit{ifClosest}(b'', b') = \begin{cases} 1, & \text{if } b' = \arg\min_{b_n} |b'' - b_n| \\ 0, & \text{otherwise} , \end{cases}$$

where $b_n$ is one of the $B$ expert beliefs that appeared within the expert trajectories.

$P^\pi(b, b')$ is an approximate belief state transition model. It is approximated in three steps. First, the next belief $b''$ is estimated using the SE function. Second, the *ifClosest* function is used to find, $b'$, the nearest belief that occurred within the expert trajectories. Finally, the transition probability between $b$ and $b'$ is updated using Eq. (5.12). This avoids handling the excessive number of new beliefs created by the SE function. More importantly, this procedure supports the use of IRL on a fixed number of beliefs, such as expert beliefs from a fixed number of trajectories.

Figure 5.2 demonstrates how the belief transition matrix is constructed for a candidate policy $\pi$. Assume that the expert beliefs include only two belief points: $b_0$ and $b_1$, as shown in Fig. 5.2 top left. Then, the belief transition matrix is initialized to zero, as shown in Fig. 5.2 top right. Starting from belief $b_1$, the action $\pi(b_1)$ is taken. If the observation $o_1$ is received then, using *SE* function, the new belief $\tilde{b}_1$ is created, shown in Fig. 5.2 middle left. The closest expert belief to $\tilde{b}_1$ is $b_0$, so the probability $\Pr(o_1|b_1, \pi(b_1))$ is added to the transition from $b_1$ (the starting belief) to $b_0$ the landed belief, as shown in Fig. 5.2 middle right. On the other hand, if the observation $o_2$ is received, then, using *SE* function, the new belief $\tilde{b}_2$ is created, shown in Fig. 5.2 bottom left. The closest expert belief to $\tilde{b}_2$ is $b_1$, so the probability $\Pr(o_2|b_1, \pi(b_1))$ is added to the transition from $b_1$ (the starting belief) to $b_1$ the landed belief, as shown in Fig. 5.2 bottom right.

We construct the rest of formulations similar to MDPs. The reward function, $R$, is represented using the vector of features $\boldsymbol{\phi}$ so that each $\phi_i(s, a)$ is a basis function for the reward function. However, in POMDPs, we need to extend state features to beliefs. To do so, we define the vector $\phi(b, a)$ as: $\boldsymbol{\phi}(b, a) = \sum_{s \in S} b(s)\boldsymbol{\phi}(s, a)$. Then, matrix $\boldsymbol{\Phi}_b^\pi$ is an $N \times K$ matrix of belief action features for policy $\pi$, defined as:

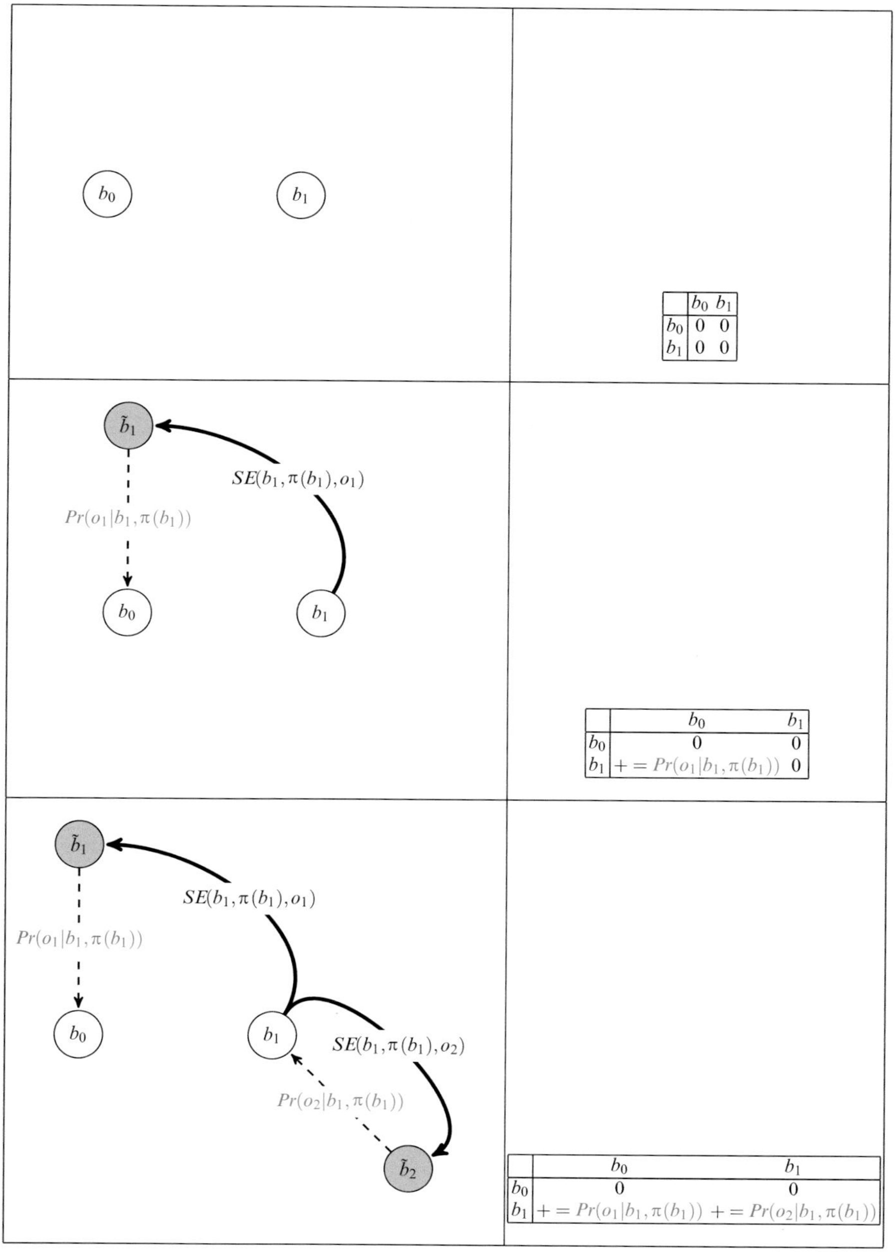


**Fig. 5.2** POMDP-IRL-BT illustration example

$$\boldsymbol{\Phi}_b^{\pi} = \begin{pmatrix} \boldsymbol{\phi}(b_0, \pi(b_0))^{\mathrm{T}} \\ \dots \\ \boldsymbol{\phi}(b_{B-1}, \pi(b_{B-1}))^{\mathrm{T}} \end{pmatrix}.$$

For the expert policy $\pi_E$, we define $\boldsymbol{\Phi}_b^{\pi_E}$ as:

$$\boldsymbol{\Phi}_b^{\pi_E} = \begin{pmatrix} \boldsymbol{\phi}(b_0, \pi_E(b_0))^{\mathrm{T}} \\ \dots \\ \boldsymbol{\phi}(b_{B-1}, \pi_E(b_{B-1}))^{\mathrm{T}} \end{pmatrix}.$$

Formally, we define $\boldsymbol{r}_b^{\pi}$ as:

$$\boldsymbol{r}_b^{\pi} = \boldsymbol{\Phi}_b^{\pi} \boldsymbol{\alpha}. \tag{5.13}$$

Similar to the MDP-IRL, we can manipulate Eq. (5.11):

$$\begin{aligned} \boldsymbol{v}_b^{\pi} &= \boldsymbol{r}_b^{\pi} + \gamma \boldsymbol{P}^{\pi} \boldsymbol{v}_b^{\pi} \\ \boldsymbol{v}_b^{\pi} - \gamma \boldsymbol{P}^{\pi} \boldsymbol{v}_b^{\pi} &= \boldsymbol{r}_b^{\pi} \\ (\boldsymbol{I} - \gamma \boldsymbol{P}^{\pi}) \boldsymbol{v}_b^{\pi} &= \boldsymbol{r}_b^{\pi} \\ \boldsymbol{v}_b^{\pi} &= (\boldsymbol{I} - \gamma \boldsymbol{P}^{\pi})^{-1} \boldsymbol{r}_b^{\pi}. \end{aligned}$$

Therefore, from the last equality we have:

$$\boldsymbol{v}_b^{\pi} = (\boldsymbol{I} - \gamma \boldsymbol{P}^{\pi})^{-1} \boldsymbol{r}_b^{\pi}. \tag{5.14}$$

Using Eq. (5.13) in Eq. (5.14), we have:

$$\begin{aligned} \boldsymbol{v}_b^{\pi} &= (\boldsymbol{I} - \gamma \boldsymbol{P}^{\pi})^{-1} \boldsymbol{\Phi}_b^{\pi} \boldsymbol{\alpha} \qquad (5.15) \\ \boldsymbol{v}_b^{\pi} &= \boldsymbol{x}_b^{\pi} \boldsymbol{\alpha}, \end{aligned}$$

where $\boldsymbol{x}_b^{\pi}$ is a matrix of size $B \times K$ defined as:

$$\boldsymbol{x}_b^{\pi} = (\boldsymbol{I} - \gamma \boldsymbol{P}^{\pi})^{-1} \boldsymbol{\Phi}_b^{\pi}. \tag{5.16}$$

Equation (5.15) shows that the vector of values $\boldsymbol{v}_b^{\pi}$ can be represented as multiplication of the vector of feature weights $\boldsymbol{\alpha}$ and the vector $\boldsymbol{x}^{\pi_b}$.

We have a similar equation for the expert policy: $\boldsymbol{v}_b^{\pi_E} = \boldsymbol{x}_b^{\pi_E} \boldsymbol{\alpha}$, where $\boldsymbol{x}_b^{\pi_E}$ is a matrix of size $B \times K$ defined as:

$$\boldsymbol{x}_b^{\pi_E} = (\boldsymbol{I} - \gamma \boldsymbol{P}^{\pi_E})^{-1} \boldsymbol{\Phi}_b^{\pi_E}, \tag{5.17}$$

where $\boldsymbol{P}^{\pi_E}$ is an $B \times B$ matrix where each element $\boldsymbol{P}^{\pi_E}(b_i, b_j)$ is the probability of transiting from $b_i$ to $b_j$ with expert action $\pi_E(b_i)$.

Algorithm 8 shows POMDP-IRL-BT. Similar to MDP-IRL, this algorithm maximizes the sum of the margins between the expert policy $\pi_E$ and the candidate policies $\pi_t$(Line 7 of Algorithm 8). The POMDP-IRL-BT algorithm is based on the belief transition model, as opposed to MDP-IRL which is based on transition of completely observable states.

Let's go through Algorithm 8 in detail. The algorithm starts by randomly initiating values for $\boldsymbol{\alpha}$ to generate the initial candidate reward function $R^1$ in Line 1. Then, the algorithm finds the policy of $R^1$, denoted by $\pi_1$, using a model-based POMDP algorithm such as point-based value iteration (PBVI) (Pineau et al. 2003). In Line 3, $\boldsymbol{P}^{\pi_1}$ is constructed, which is used to calculate $\boldsymbol{x}_b^{\pi_1}$ from Eq. (5.16). Then, in Line 4, the expert policy $\pi_E$ is used to construct $\boldsymbol{P}^{\pi_E}$ which is used to calculate $\boldsymbol{x}_b^{\pi_E}$ from Eq. (5.17).

---

**Algorithm 8:** POMDP-IRL-BT: inverse reinforcement learning in the POMDP framework using belief transition estimation

---

**Input**: $POMDP \backslash R = \{S, A, T, \gamma, O, \Omega, b_0\}$, expert trajectories in the form of $D = \{(b_n, \pi_E(b_n), b_n')\}$, a vector of features $\boldsymbol{\phi} = (\phi_1, \ldots, \phi_K)$, convergence rate $\epsilon$, and maximum Iteration $maxT$

**Output**: Finds reward function $R$ where $R = \sum_i \alpha_i \phi_i(s, a)$ by approximating $\boldsymbol{\alpha} = (\alpha_1, \ldots, \alpha_K)$

1 Choose the initial reward $R^1$ by randomly initializing feature weights $\boldsymbol{\alpha}$;
2 Set $\Pi = \{\pi_1\}$ by finding $\pi_1$ using POMDP with candidate reward function $R^1$ and a PBVI variant POMDP solver;
3 Set $X = \{\boldsymbol{x}_b^{\pi_1}\}$ by calculating $\boldsymbol{x}_b^{\pi_1}$ using $\boldsymbol{P}^{\pi_1}$ and Equation (5.16);
4 Calculate $\boldsymbol{x}_b^*$ from Equation (5.17);
5 **for** $t \leftarrow 1$ *to* $maxT$ **do**
6 Find values for $\boldsymbol{\alpha}$ by solving the linear program:
7 $\quad$ maximize$_{\boldsymbol{\alpha}}$ $\left[ \left( (\boldsymbol{x}_b^{\pi_E} - \boldsymbol{x}_b^{\pi_1}) + \ldots + (\boldsymbol{x}_b^{\pi_E} - \boldsymbol{x}_b^{\pi_t}) \right) \boldsymbol{\alpha} \right]$;
8 $\quad$ subject to $-1 \le \alpha_i \le +1 \; \forall i \; 1 \le i \le K$;
9 $R^{t+1} = \sum_i \alpha_i^t \phi_i(s, a)$;
10 **if** $max_i |\boldsymbol{\alpha}_i^t - \boldsymbol{\alpha}_i^{t-1}| \le \epsilon$ **then**
11 $\quad$ return $R^{t+1}$;
12 **end**
13 **else**
14 $\quad$ $\Pi = \Pi \cup \{\pi_{t+1}\}$ by finding $\pi_{t+1}$ using POMDP with candidate reward function $R^{t+1}$ and a PBVI variant POMDP solver;
15 $\quad$ Set $X = X \cup \{\boldsymbol{x}_b^{\pi_{t+1}}\}$ by calculating $\boldsymbol{x}_b^{\pi_{t+1}}$ using $\boldsymbol{P}^{\pi_{t+1}}$ and Equation (5.16);
16 **end**
17 **end**

---

From Line 5 to Line 17, POMDP-IRL-BT iterates to learn the expert reward function by solving the linear program in Line 7 with the constraints shown in Line 8. The objective function of the linear program is:

$$\text{maximize}_{\boldsymbol{\alpha}}\ \boldsymbol{d}^t = \sum_{l=1}^{t} (\boldsymbol{x}_b^{\pi_E}\boldsymbol{\alpha} - \boldsymbol{x}_b^{\pi_l}\boldsymbol{\alpha})$$

for all $t$ candidate policies learned so far up to iteration $t$, subject to the constraints $|\alpha_i| \leq 1\ \forall i\ 1 \leq i \leq K$. So, it maximizes the sum of the margins between expert policy $\pi^*$ and other candidate policies $\pi_l$ (we have $t$ of them at iteration $t$). The rest is similar to the MDP-IRL. In this optimization, we also constrain the value of the expert's policy to be greater than that of other policies in order to ensure that the expert's policy is optimal.

As seen above, POMDP-IRL-BT approximates the expert policy value and the candidate policy values in POMDPs using the belief transition that is approximated in Eq. (5.12). This approximation is done by first fixing the number of beliefs to expert beliefs. Moreover, after performing action $a$ in a belief, we may end up to a new belief $b''$ (outside expert beliefs) which we map it to the closest expert belief.

In our previous work (Chinaei and Chaib-draa 2012), we applied the POMDP-IRL-BT algorithm on POMDP benchmarks. Furthermore, we applied the algorithm on the dialog POMDP learned from SmartWheeler (described in Chap. 6). The experimental results showed that the algorithm is able to learn a reward function that accounts for the expert policy. In Chap. 6, we apply the proposed methods in this book to learn a dialog POMDP from SmartWheeler dialogs; we also apply POMDP-IRL-BT on the learned dialog POMDP and demonstrate the results.

### 5.3.2 PB-POMDP-IRL

In this section, we propose a point-based IRL algorithm for POMDPs, called PB-POMDP-IRL. The idea in this algorithm is that the value of new beliefs, i.e., the beliefs that are result of performing other policies than expert policy, are approximated using expert beliefs. Moreover, this algorithm constructs a linear program for learning a reward function for the expert policy by going through the expert trajectories and adding variables corresponding to the expert policy value and variables corresponding to the alternative policy values.

To understand the algorithm, we start by some definitions: we define each history $\boldsymbol{h}$ as a sequence of observation action pairs of the expert trajectories denoted by $\boldsymbol{h} = ((a_1, o_1), \ldots, (a_t, o_t))$. Moreover, we use $\boldsymbol{hao}$ for the history of size $|\boldsymbol{h}| + 1$ which includes the history $\boldsymbol{h}$ followed by $(a, o)$. Then, we use $b_{\boldsymbol{h}}$ to show the belief at the end of history $\boldsymbol{h}$, which can be calculated using the State Estimator in Eq. (3.3). We present State Estimator function again here:

$$\begin{aligned} b_{\boldsymbol{h}ao}(s') &= \mathrm{SE}(b_{\boldsymbol{h}}, a, o) \\ &= \Pr(s'|b_{\boldsymbol{h}}, a, o) \\ &= \eta \Omega(a, s', o) \sum_{s \in S} b_{\boldsymbol{h}}(s) T(s, a, s'), \end{aligned}$$

where $\eta$ is the normalization factor.

For instance, if $\boldsymbol{h} = (a_1, o_1)$, then the belief at the end of history $\boldsymbol{h}$, $b_{\boldsymbol{h}}$ is calculated by the belief update function in Eq. (3.3) and using $(a_1, o_1)$ and $b_0$ (usually a uniform belief) as the parameters. Similarly, if $\boldsymbol{h} = ((a_1, o_1), \ldots, (a_t, o_t,))$, the belief at the end of history is calculated by sequentially applying the belief update using $(a_i, o_i)$ and $b_{i-1}$ as the parameters.

The PB-POMDP-IRL algorithm is described in Algorithm 9. In our proposed algorithm, the value of new beliefs, i.e., the beliefs which are result of performing other policies (than expert policy), are approximated using expert beliefs. That is, given the belief $b_{\boldsymbol{h}ao}$ where $a \neq \pi_E(b_{\boldsymbol{h}ao})$, the value of $V^{\pi_E}(b_{\boldsymbol{h}ao})$ is approximated using expert histories $\boldsymbol{h}'_i$ of the same size as $\boldsymbol{h}ao$, i.e., $|\boldsymbol{h}'_i| = |\boldsymbol{h}ao|$. This approximation is demonstrated in Line 15 and Line 16 of the algorithm:

$$V^{\pi_E}(b_{\boldsymbol{h}ao}) = \sum_{i=0}^{n} w_i V(b_{\boldsymbol{h}'_i})$$

such that $w_i$s follow:

$$b_{\boldsymbol{h}ao} = \sum_{i=0}^{n} w_i b_{\boldsymbol{h}'_i}.$$

Notice that due to the piecewise linearity of the optimal value function, this approximation corresponds to the true value if the expert policy in the belief state $b_{\boldsymbol{h}ao}$ is the same as the one in the belief states $b_{\boldsymbol{h}'_i}$, which is used in the linear combination. This condition is more likely to be true when the beliefs $b_{\boldsymbol{h}'_i}$ are closer to the approximated belief $b_{\boldsymbol{h}ao}$.

**Algorithm 9:** Point-based POMDP-IRL: a point-based algorithm for IRL in the POMDP framework

**Input**: A POMDP$\backslash R$ as $(S, A, O, T, \Omega, b_0, \gamma)$, expert trajectories $D$ in the form of $a_1^m o_1^m \ldots a_{t-1}^m o_{t-1}^m a_t^m, t \leq H$

**Output**: Reward weights $\alpha_i \in R$;

Extract the human's policy $\pi_E$ from the trajectories;
Initialize the set of variables $V$ with the weights $\alpha_i$;
Initialize the set of linear constraints $C$ with $\{\forall (s,a) \in S \times A : R_{min} \leq \sum_{i=1}^{k} \alpha_i \phi_i(s,a) \leq R_{max}\}$ ;
**for** $t \leftarrow H$ **to** $1$ **do**
  **foreach** $\boldsymbol{h} \in D$, *such that* $\boldsymbol{h}$ *is a trajectory of length* $t$, **do**
    Calculate $b_{\boldsymbol{h}}$, the belief state at the end of trajectory $\boldsymbol{h}$;
    **foreach** $(a, o) \in A \times O$ **do**
      Add the variable $V^{\pi_E}(b_{\boldsymbol{h}ao})$ to $V$;
      /* $V^{\pi_E}(b_{\boldsymbol{h}ao})$ is approximation of $\pi^{\pi_E}$ value at $b_{\boldsymbol{h}ao}$ defined below */
      **if** $\boldsymbol{h}ao \notin D$ *and* $t = H$ **then**
        Add the constraint $V^{\pi_E}(b_{\boldsymbol{h}ao}) = 0$ to the set $C$ ;
      **end**
      **if** $\boldsymbol{h}ao \notin D$ *and* $t < H$ **then**
        Let $b_{\boldsymbol{h}ao}$ be the belief corresponding to the trajectory $\boldsymbol{h}ao$;
        Calculate the belief states $b_{\boldsymbol{h}'_i}$ corresponding to the trajectories in $D$ of length $t+1$ ;
        Find a list of weights $w_i$ such that $b_{\boldsymbol{h}ao} = \sum_{i=0}^{n} w_i b_{\boldsymbol{h}'_i}$;
        Add to $C$ the constraint $V^{\pi_E}(b_{\boldsymbol{h}ao}) = \sum_{i=0}^{n} w_i V(b_{\boldsymbol{h}'_i})$;
        /* $V^{\pi_E}(b_{\boldsymbol{h}ao})$ is approximation of $\pi_E$ value at the belief corresponding to the trajectory $\boldsymbol{h}ao$ */
      **end**
    **end**
    Add the variable $V^{\pi_E}(b_{\boldsymbol{h}})$ to $V$;
    /* $V^{\pi_E}(b_{\boldsymbol{h}})$ is $\pi^{\pi_E}$ value at $b_{\boldsymbol{h}}$ */
    Add to $C$ the constraint $V^{\pi_E}(b_{\boldsymbol{h}}) = \left[\sum_{s \in S} b_{\boldsymbol{h}}(s) \sum_{i=1}^{k} \alpha_i \phi_i(s, \pi_E(b_{\boldsymbol{h}})) + \gamma \sum_{o \in O} Pr(o|b_{\boldsymbol{h}}, \pi_E(b_{\boldsymbol{h}})) V^{\pi_E}(b_{\boldsymbol{h}\pi_E(b_{\boldsymbol{h}})o})\right]$ ;
    **foreach** $a \in A$ **do**
      Add the variable $V^a(b_{\boldsymbol{h}})$ to $V$;
      /* $V^a(b_{\boldsymbol{h}})$ is the value of the alternative policy that chooses $a$ after the trajectory $\boldsymbol{h}$ */
      Add to $C$ the constraint $V^a(b_{\boldsymbol{h}}) = \sum_{s \in S} b_{\boldsymbol{h}}(s) \sum_{i=1}^{k} \alpha_i \phi_i(s,a) + \gamma \sum_{o \in O} Pr(o|b_{\boldsymbol{h}}, a) V^{\pi_E}(b_{\boldsymbol{h}ao})$;
      Add the variable $\epsilon_{\boldsymbol{h}}^a$ to the set $V$;
      Add to $C$ the constraint $V^{\pi_E}(b_{\boldsymbol{h}}) - V^a(b_{\boldsymbol{h}}) \geq \epsilon_{\boldsymbol{h}}^a$;
    **end**
  **end**
**end**
maximize $\sum_{\boldsymbol{h} \in H} \sum_{a \in A} \epsilon_{\boldsymbol{h}}^a$ subject to the constraints of set $C$;

The algorithm also constructs a linear program for learning the reward function by going through expert trajectories and adding variables corresponding to the expert policy value and variables corresponding to alternative policy values. These variables are subject to the linear constraints that are subject to the Bellman equation (Line 20 and Line 23). In Line 20, the linear constraint for the expert policy value at end of history $\boldsymbol{h}$ is added. This constraint is based on the Bellman equation (3.5) which we present it again here:

$$V^{\pi}(b) = R(b, \pi(b)) + \gamma \sum_{o' \in O} \Pr(o'|b, \pi(b)) V^{\pi}(b'),$$

where here the rewards are presented as linear combination of state features:

$$R(s, a) = \sum_{i=1}^{k} \alpha_i \phi_i(s, a)$$

and $R(b, a)$ is defined as:

$$\sum_{s \in S} b(s) R(s, a).$$

So, the value of expert policy at end of history $\boldsymbol{h}$ becomes:

$$V^{\pi_E}(b_{\boldsymbol{h}}) = \left[ \sum_{s \in S} b_{\boldsymbol{h}}(s) \sum_{i=1}^{k} \alpha_i \phi_i(s, \pi^{\pi_E}(b_{\boldsymbol{h}})) + \gamma \sum_{o \in O} \Pr(o|b_{\boldsymbol{h}}, \pi^{\pi_E}(b_{\boldsymbol{h}})) V^{\pi_E}(b_{\boldsymbol{h}\pi^{\pi_E}(b_{\boldsymbol{h}})o}) \right].$$

Similarly, in Line 23 the linear constraint for the alternative policy value at the end of history $\boldsymbol{h}$ is added. Notice that an alternative policy is a policy that selects an action $a \neq \pi^{\pi_E}(b_{\boldsymbol{h}})$ and then follows the expert's policy for the upcoming time-steps. This constraint is also based on the Bellman equation (3.5). That is, the value of performing action $a$ at the belief $b_{\boldsymbol{h}}$ where $a \neq \pi^{\pi_E}(b_{\boldsymbol{h}})$ and then following expert policy $\pi_E$ becomes:

$$V^{a}(b_{\boldsymbol{h}}) = \sum_{s \in S} b_{\boldsymbol{h}}(s) \sum_{i=1}^{k} \alpha_i \phi_i(s, a) + \gamma \sum_{o \in O} \Pr(o|b_{\boldsymbol{h}}, a) V^{\pi_E}(b_{\boldsymbol{h}ao}).$$

Finally, in Line 25 we explicitly state that the expert policy value at any history $\boldsymbol{h}$, $V^{\pi_E}(b_{\boldsymbol{h}})$ is higher than any alternative policy value, $V^{a}(b_{\boldsymbol{h}})$ where $a \neq \pi_E(b_{\boldsymbol{h}})$, by a margin $\epsilon_{\boldsymbol{h}}^{a}$ that should be maximized in Line 29.

### 5.3.3 PB-POMDP-IRL Evaluation

In our previous work (Boularias et al. 2010), we evaluated the PB-POMDP-IRL performance as the ASR noise level increases. The results are shown in Table 5.1. We applied the algorithm on four dialog POMDPs learned from SACTI-1 dialogs

**Table 5.1** Number of matches for hand-crafted reward POMDPs, and learned reward POMDPs, w.r.t. 1415 human expert actions

| Noise level | None | Low | Med | High |
|---|---|---|---|---|
| HC reward matches | 339-24 % | 327-23 % | 375-26 % | 669-47 % |
| Learned reward matches | **869-61 %** | **869-61 %** | **408-28 %** | **387-27 %** |

with four levels of noise *none*, *low*, *medium*, and *high*, respectively, as described in Sect. 4.5.2. Our experimental results showed that the PB-POMDP-IRL algorithm is able to learn a reward function for *human* expert policy. The results in Table 5.1 shows that the actions suggested by the learned reward match to the human actions highly better than the actions suggested by the assumed reward function. This is highlighted in Table 5.1. Note that SACTI dialogs have been collected in a Wizard-of-Oz setting. The results also show that the algorithm performs better in the lower noise levels (*none* and *low*) than in higher noise levels (*medium* and *high*). In Sect. 5.6, we compare the PB-POMDP-IRL algorithm to the POMDP-IRL-BT algorithm on SmartWheeler learned POMDP actions.

## 5.4 Related Work

IRL has been mostly developed in the MDP framework. In particular, in Sect. 5.2, we studied the basic trajectory-based MDP-IRL algorithm, proposed by Ng and Russell (2000). Later on, Abbeel and Ng (2004) introduced an apprenticeship learning algorithm via IRL, which aims to find a policy which is *close to* the expert policy. That is, a policy whose feature expectations are close to that of expert policy. The feature expectations are derived from the MDP value function in Eq. (3.1), which we present it again here:

$$\begin{aligned} V^{\pi}(s) &= E_{s\sim T}\left[\sum_{t=0}^{\infty}\gamma^t R(s,\pi(s))|s_0\right] \\ &= E_{s\sim T}\left[\sum_{t=0}^{\infty}\gamma^t \boldsymbol{\alpha}^{\mathrm{T}}\boldsymbol{\phi}(s,\pi(s))|s_0\right] \\ &= \boldsymbol{\alpha}^{\mathrm{T}}E_{s\sim T}\left[\sum_{t=0}^{\infty}\gamma^t \boldsymbol{\phi}(s,\pi(s))|s_0\right] \\ &= \boldsymbol{\alpha}^{\mathrm{T}}\mu(\pi), \end{aligned} \tag{5.18}$$

where the second equality is because the reward function is represented as the linear combination of features, similar to MDP-IRL, we have $R(s,a) = \alpha\phi(s,a)$.

From Eq. (5.18), we can see:

$$\mu(\pi) = E_{s\sim T}\left[\sum_{t=0}^{\infty} \gamma^t \boldsymbol{\phi}(s, \pi(s))|s_0\right]$$

in which $\mu(\pi)$ is the vector of expected discounted feature values $\mu(\pi)$, i.e., feature expectations. By comparing the definition of feature expectation $\mu(\pi)$ to the vector $\boldsymbol{x}^{\pi}$ appearing in Eq. (5.5), we learn that the vector $\boldsymbol{x}^{\pi}$ is an approximation for feature expectation.

Then, the apprenticeship learning problem is reduced to the problem of finding a policy whose feature expectation is close to the expert policy feature expectation. This is done by learning a reward function as an intermediate step. Notice that in apprenticeship learning the learned reward function is *not* necessarily the correct underlying reward function (Abbeel and Ng 2004); as the objective in the algorithm is finding the reward function for the policy that has an approximate feature expectation close to the expert policy feature expectation.

In the POMDP framework, as mentioned in Sect. 5.1, Choi and Kim (2011) provided a general framework for IRL in POMDPs by assuming that expert policy is represented in the form of an FSC, and thus using an FSC-based POMDP solver called PBPI (point-based policy iteration) (Ji et al. 2007). Similar to the trajectory-based algorithms introduced in this chapter, Choi and Kim (2011) proposed trajectory-based algorithms for learning the POMDP reward functions (besides their proposed analytical-based algorithms). In particular, they proposed a trajectory-based algorithm called MMV (max-margin between values) described as follows.

The MMV algorithm is similar to the MDP-IRL algorithm, introduced in Sect. 5.2, which works given the MDP model and expert trajectories. In particular, Choi and Kim (2011) used an objective function for maximizing the sum of the margin between expert policy and other candidate policies using a monotonic function $f$, similar to Ng and Russell (2000) (cf. end of Sect. 5.2). Moreover, the policy values are estimated using the Monte Carlo estimator using expert trajectories. Recall the definition of value function in POMDPs, shown in Eq. (3.5), defined as:

$$\begin{aligned} V^{\pi}(b) &= E_{b_t\sim \text{SE}}\left[\gamma^0 R(b_0, \pi(b_0)) + \gamma^1 R(b_1, \pi(b_1) + \cdots)|\pi, b_0 = b\right] \\ &= E_{b_t\sim \text{SE}}\left[\sum_{t=0}^{\infty} \gamma^t R(b_t, \pi(b_t))|\pi, b_0 = b\right]. \end{aligned}$$

Using an expert trajectory of size $B$, the value of expert policy can be estimated using the Monte Carlo estimator as:

$$\begin{aligned} \hat{V}^{\pi_E}(b_0) &= R(b_0, \pi_E(b_0)) + \cdots + R(b_{B-1}, \pi(b_{B-1})) \\ &= \sum_{t=0}^{B-1} \gamma^t R(b_t, a_t) \end{aligned} \tag{5.19}$$

$$= \boldsymbol{\alpha}^{\mathrm{T}} \sum_{t=0}^{B-1} \gamma^t \boldsymbol{\phi}(b_t, a_t),$$

where the last equality comes from the reward function representation using features, shown in Eq. (5.13).

Similar to the trajectory-based MMV algorithm of Choi and Kim (2011), we used the POMDP beliefs that appeared in the expert trajectories. In contrast to the FSC-based representation used in Choi and Kim (2011), we used the belief point representation. Furthermore, instead of approximating the policy values using the Monte Carlo estimator, we approximated the policy values by approximating the belief transition matrix in Eq. (5.12).

In order to compare the belief transition estimation to the Monte Carlo estimation, we implemented the Monte Carlo estimator in the POMDP-IRL-BT algorithm. This new algorithm is called POMDP-IRL-MC (MC for the Monte Carlo estimator) and described as follows.

## 5.5 POMDP-IRL-MC

Estimating policy values can be inaccurate, in both the introduced methods: the Monte Carlo estimator and the belief transition approximation, proposed in Eq. (5.12) (in the POMDP-IRL-BT algorithm). This is because the number of expert trajectories is small compared to the infinite number of possible belief points. In order to compare the Monte Carlo estimation to the belief transition estimation, we implemented the Monte Carlo estimator in Eq. (5.19) for estimation of policy values in Line 7 of Algorithm 8, and used the Perseus software (Spaan and Vlassis 2005) as the POMDP solver. This new algorithm is called POMDP-IRL-MC which is similar to the MMV algorithm of Choi and Kim (2011), described in the previous section.

The difference between the MMV algorithm of Choi and Kim (2011) and POMDP-IRL-MC is the policy representation and consequently the POMDP solver. As mentioned above, Choi and Kim (2011) used FSC representation in their MMV algorithm and thus using PBPI, an FSC-based POMDP solver (Ji et al. 2007). In POMDP-IRL-MC, however, we used belief point representation and thus used, Perseus, a point-based POMDP solver (Spaan and Vlassis 2005) (similar to our POMDP-IRL-BT algorithm, proposed in Sect. 5.3.1). In Sect. 6.3.4, we compare the POMDP-IRL-BT algorithm to the POMDP-IRL-MC in terms of solution quality and scalability.

**Table 5.2** The learned SACTI-1 specification for IRL experiments

| Problem | $\|S\|$ | $\|A\|$ | $\|O\|$ | $\gamma$ | $\|\phi\|$ | \|Trajectories\| |
|---|---|---|---|---|---|---|
| SACTI-1 | 5 | 14 | 5 | 0.90 | 70 | 50 |

## 5.6 POMDP-IRL-BT and PB-POMDP-IRL Performance

In this section, we show the example of IRL on the learned dialog POMDP from SACTI-1, introduced in Sect. 4.5. In particular, we apply POMDP-IRL-BT (introduced in Sect. 5.3.1) and PB-POMDP-IRL (introduced in Sect. 5.3.2) for learning the reward function of our example dialog POMDP learned from SACTI-1. Recall the learned intent dialog POMDP from SACTI-1. The POMDP model consists of five states, three non-terminal states for *visits*, *transports*, and *foods* intents, as well as two terminal states *success* and *failure*. The POMDP model also includes 14 actions, 5 intent observations, and the learned transition and observation models. The learned SACTI-1 specification for IRL experiments, of this section, is described in Table 5.2.

As mentioned in Sect. 5.1, for the purpose of POMDP-IRL experiments, we consider expert policy as a POMDP policy similar to the previous works (Ng and Russell 2000; Choi and Kim 2011). For the expert reward function, we assumed the reward function introduced in Sect. 4.5. That is, the reward function which penalizes each action in non-terminal states by $-1$. Moreover, any action in the *success* terminal state receives +50 as reward, and any action in the *failure* terminal state receives $-50$ as reward. Then, we solved the POMDP model to find the optimal policy and assumed it as the expert policy to generate ten trajectories. Each trajectory is generated from the initial belief and by performing the expert action. After receiving an observation the expert belief is updated and the next action is performed. The trajectory ends when reaching one of the two terminal states. The ten generated trajectories were then used in our two fold cross validation experiments.

We applied the POMDP-IRL-BT and PB-POMDP-IRL algorithms on the SACTI-1 dialog POMDP using *state-action-wise* features in which there is an indicator function for each state-action pair. Since there are 5 states and 14 actions in the example dialog POMDP, the size of features equals $70 = 5 \times 14$. To solve each POMDP model, we used the Perseus solver which is a PBVI solver (Spaan and Vlassis 2005). As stated in Sect. 3.1.4.4, PBVI solvers are approximate solvers that use a finite number of beliefs for solving a POMDP model. We set the solver to use 10,000 random samples for solving the optimal policy of each candidate reward. The other parameter is max-time for execution of the algorithm, which is set to 1000.

The two fold cross validation experiments are done as follows. We randomly selected five trajectories from the ten expert trajectories, introduced above, for training and the rest of five trajectories for testing. Then we tested POMDP-IRL-BT and PB-POMDP-IRL. For each algorithm experiment, the algorithm was used to learn a reward function for the expert trajectories using the training trajectories. Then the learned policy, i.e., the policy of the learned reward function, was applied

**Table 5.3** POMDP-IRL-BT and PB-POMDP-IRL results on the learned POMDP from SACTI-1: number of matched actions to the expert actions

| Algorithm | # of matched actions | Matched-percentage |
|---|---|---|
| POMDP-IRL-BT | 42 | 84 % |
| PB-POMDP-IRL | 29 | 58 % |

on the testing trajectories. Finally, we calculated the number of learned actions that matched to the expert actions on the testing trajectories, and they were added up for the two folds to make the cross validation experiments complete.

The experimental results are shown in Table 5.3. The results show that POMDP-IRL-BT significantly outperforms PB-POMDP-IRL. More specifically, the POMDP-IRL-BT algorithm was able to learn a reward function that matched with 42 actions out of 50 actions in the data set. That is, the policy of the learned reward function was equal to the expert policy for 84 % of the beliefs. On the other hand, the learned policy using PB-POMDP-IRL matched to 29 actions out of the 50 actions in the data set, i.e., 58 % match. Thus, in the next chapter, for learning the reward function, we apply POMDP-IRL-BT on the learned dialog POMDP from SmartWheeler.

## 5.7 Conclusions

In this chapter, we first introduced IRL for learning the reward function of expert policy in the MDP framework. In particular, we studied MDP-IRL algorithm of Ng and Russell (2000), the basic IRL algorithm in the MDP framework. Then, we proposed two IRL algorithms in the POMDP framework: POMDP-IRL-BT and PB-POMDP-IRL.

The proposed POMDP-IRL-BT algorithm is similar to the MDP-IRL algorithm. That is, it maximizes sum of the margin between the expert policy and other intermediate candidate policies. Moreover, instead of states we used belief states and the optimization is performed only on the expert beliefs, rather than all possible beliefs, using an approximated belief transition model. On the other hand, the idea in the proposed PB-POMDP-IRL algorithm is that the value of new beliefs, i.e., the beliefs that are result of performing other policies than expert policy, are linearly approximated using expert belief values. We then revisited the learned intent POMDP from SACTI-1 and applied the two proposed POMDP-IRL algorithms on it. The result of the experiments showed that POMDP-IRL-BT significantly outperforms PB-POMDP-IRL.

Learning the reward function from expert dialogs makes our descriptive Algorithm 1 complete. In the following chapter, we show the application of our proposed methods on healthcare dialog management.

# Chapter 6
# Application on Healthcare Dialog Management

## 6.1 Introduction

In this chapter, we show the application of our proposed methods on healthcare dialog management (Chinaei et al. 2014). That is, we use the methods in this book to learn a dialog POMDP from real dialogs of an intent-based dialog domain (cf. Chap. 1), known as SmartWheeler (Pineau et al. 2011). The SmartWheeler project aims to build an intelligent wheelchair using for persons with disabilities. In recent years, there has been extensive work on machine learning-based robotic systems (Cuayáhuitl et al. 2013). In particular, SmartWheeler aims to minimize the physical and cognitive load required in steering it. This project has been initiated in 2006, and a first prototype, shown in Fig. 6.1, was built in-house at McGill's Center for Intelligent Machines.

We used the dialogs collected by SmartWheeler to develop a dialog POMDP learned primarily from data. The data includes eight dialogs with healthy users and nine dialogs with target users of SmartWheeler (Pineau et al. 2011). The dialogs with target users, who are the elderly, are somehow more noisy than the ones with healthy users. More specifically, the average word error rate (WER) equals 13.9 % for the healthy user dialogs and 18.5 % for the target user dialogs. In order to perform our experiments on a larger amount of data, we used all the healthy and target user dialogs. In total, there are 2853 user utterances and 422 distinct words in the SmartWheeler dialogs.

Table 6.1 shows a sample of SmartWheeler dialogs captured for training the dialog POMDP model components. The first line denoted by $u_1$ shows the true user utterance, that is the one which has been extracted manually from user audio recordings. The following line denoted by $\tilde{u}_1$ is the recognized user utterances by ASR. Finally, the line denoted by $a_1$ shows the performed action in response to the ASR output at the time of collecting the dialogs.



H. Chinaei, B. Chaib-draa, *Building Dialogue POMDPs from Expert Dialogues*, SpringerBriefs in Speech technology, DOI 10.1007/978-3-319-26200-0_6

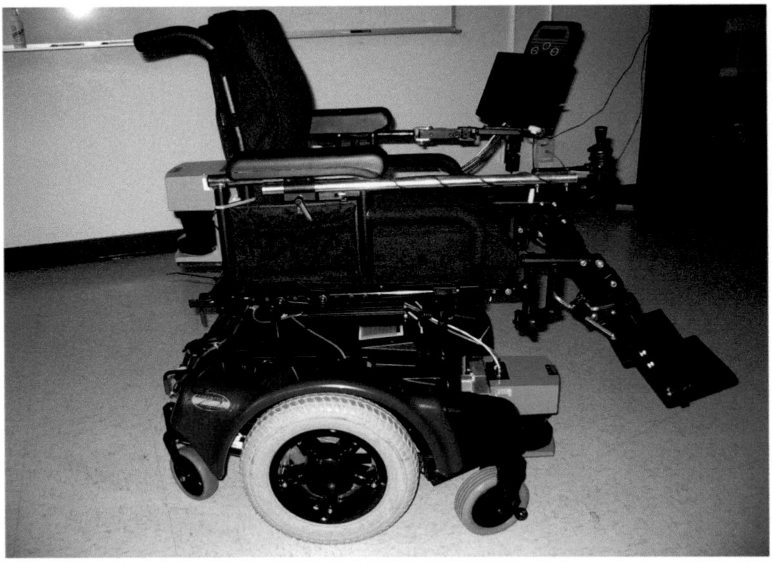

**Fig. 6.1** The SmartWheeler robot platform

**Table 6.1** A sample from the SmartWheeler dialogs (Pineau et al. 2011)

| | |
|---|---|
| | ... |
| $u_1$ : | turn right |
| $\tilde{u}_1$ : | [turn right] |
| $a_1$ : | TURN RIGHT |
| | |
| $u_2$ : | turn right a little |
| $\tilde{u}_2$ : | [10 writer little] |
| $a_2$ : | PLEASE REPEAT YOUR COMMAND |
| | |
| $u_3$ : | turn right a little |
| $\tilde{u}_3$ : | [turn right to lead a] |
| $a_3$ : | TURN RIGHT A LITTLE |
| | |
| $u_4$ : | stop |
| $\tilde{u}_4$ : | [stop] |
| $a_4$ : | STOP |
| | ... |

Notice that the true user utterance is not observable to SmartWheeler, and thus it requires to perform the action based on the recognized utterance by ASR. That is, for each dialog utterance recognized by ASR, the machine aims to estimate the user intent and then to perform the best action that satisfies the user intent.

The recognized utterance by ASR is not however reliable for decision making. For instance, the first utterance,

$$u_1 : \text{ [turn right a little]},$$

shows the true user utterance. The ASR output for this utterance is

$$\tilde{u}_1 : \text{ [10 writer little]}.$$

As such, the action performed by SmartWheeler at this dialog turn is the *general query* action

$$u_1 : \text{ PLEASE REPEAT YOUR COMMAND.}$$

The query action is the SmartWheeler action for getting more information. For instance, in the example in Table 6.1, when SmartWheeler receives the second ASR output [10 writer little], it performs a general query action to get more information before it performs the right action for the user intent, i.e., TURN RIGHT A LITTLE.

The list of all SmartWheeler actions is shown in Table 6.2. Each action is the right action of one state (the user intent for a specific command). So, ideally, there should be 24 states for SmartWheeler dialogs. (There are 24 actions other than the general query action.) However, in the next section we see that we only learned 11 of the states, mainly because of the number of dialogs. That is, not all of the states appeared in the data frequently enough. There are also states that do not appear in dialogs at all.

In this chapter, in Sect. 6.2, we learn a dialog POMDP from SmartWheeler. First in Sect. 6.2.1, we learn a keyword POMDP and an intent POMDP (without the reward function) from SmartWheeler noisy dialogs based on the introduced methods in Chap. 4. Then in Sect. 6.2.2, we compare the intent POMDP performance to the keyword POMDP performance.

In Sect. 6.3, we go through a set of experiments for IRL in SmartWheeler. First in Sect. 6.3.1, we learn a set of features for SmartWheeler, called keyword features. Then in Sect. 6.3.2, we use them for MDP-IRL application on the learned dialog MDP from SmartWheeler. Then, in Sect. 6.3.3 we experiment POMDP-IRL-BT on the SmartWheeler learned intent POMDP using the keyword features. In Sect. 6.3.4, we compare POMDP-IRL-BT and POMDP-IRL-MC, introduced in Sect. 5.5, using the learned intent POMDP from SmartWheeler. Finally, we conclude this chapter in Sect. 6.4.

## 6.2 Dialog POMDP Model Learning for SmartWheeler

We learned the possible user intents in SmartWheeler dialog based on the HTMM method as explained in Sect. 4.2.1. To do so, we preprocessed the dialogs to remove stop words such as determiners and auxiliary verbs. Then, we learned the user

**Table 6.2** The list of the possible actions, performed by SmartWheeler

| | |
|---|---|
| $a_1$ | DRIVE FORWARD A LITTLE |
| $a_2$ | DRIVE BACKWARD A LITTLE |
| $a_3$ | TURN RIGHT A LITTLE |
| $a_4$ | TURN LEFT A LITTLE |
| $a_5$ | FOLLOW THE LEFT WALL |
| $a_6$ | FOLLOW THE RIGHT WALL |
| $a_7$ | TURN RIGHT DEGREE |
| $a_8$ | GO THROUGH THE DOOR |
| $a_9$ | SET SPEED TO MEDIUM |
| $a_{10}$ | FOLLOW THE WALL |
| $a_{11}$ | STOP |
| $a_{12}$ | TURN LEFT |
| $a_{13}$ | DRIVE FORWARD |
| $a_{14}$ | APPROACH THE DOOR |
| $a_{15}$ | DRIVE BACKWARD |
| $a_{16}$ | SET SPEED TO SLOW |
| $a_{17}$ | MOVE ON SLOPE |
| $a_{18}$ | TURN AROUND |
| $a_{19}$ | PARK TO THE RIGHT |
| $a_{20}$ | TURN RIGHT |
| $a_{21}$ | DRIVE FORWARD METER |
| $a_{22}$ | PARK TO THE LEFT |
| $a_{23}$ | TURN LEFT DEGREE |
| $a_{24}$ | PLEASE REPEAT YOUR COMMAND |

intents for the SmartWheeler dialogs. Table 6.3 shows the learned user intents with their four top words. Most of the learned intents show a specific user *command*:

$$i_1 : \textit{move forward little},\ i_2 : \textit{move backward little},\ i_3 : \textit{turn right little},$$
$$i_4 : \textit{turn left little},\ i_5 : \textit{follow left wall},\ i_6 : \textit{follow right wall},$$
$$i_8 : \textit{go door}, \text{ and } i_{11} : \textit{stop}.$$

There are two learned intents that loosely represent a command:

$$i_9 : \textit{set speed} \text{ and } i_{10} : \textit{follow person}.$$

And, there is a learned intent that represents two commands:

$$i_7 : \textit{turn degree right/left}.$$

Table 6.4 shows results of HTMM application on SmartWheeler for the example shown in Table 6.1. The line denoted by $u$ is the true user utterance, manually

**Table 6.3** The learned user intents from the SmartWheeler dialogs

| | | | | | |
|---|---|---|---|---|---|
| *intent* | 1 | *intent* | 5 | *intent* | 9 |
| **forward** | 0.180 | **left** | 0.242 | **for** | 0.088 |
| move | 0.161 | wall | 0.229 | word | 0.080 |
| little | 0.114 | follow | 0.188 | speed | 0.058 |
| drive | 0.081 | fall | 0.032 | set | 0.054 |
| *intent* | 2 | *intent* | 6 | *intent* | 10 |
| **backward** | 0.380 | **right** | 0.279 | **top** | 0.143 |
| drive | 0.333 | wall | 0.212 | stop | 0.131 |
| little | 0.109 | follow | 0.197 | follow | 0.098 |
| top | 0.017 | left | 0.064 | person | 0.096 |
| *intent* | 3 | *intent* | 7 | *intent* | 11 |
| **right** | 0.209 | **turn** | 0.373 | **stop** | 0.942 |
| turn | 0.171 | degree | 0.186 | stop | 0.022 |
| little | 0.131 | right | 0.165 | scott | 0.007 |
| bit | 0.074 | left | 0.162 | but | 0.002 |
| *intent* | 4 | *intent* | 8 | | |
| **left** | 0.189 | **go** | 0.358 | | |
| turn | 0.171 | door | 0.289 | | |
| little | 0.138 | forward | 0.071 | | |
| right | 0.090 | backward | 0.065 | | |

extracted by listening to the dialog recordings. Then, $\tilde{u}$ is the recognized user utterance by ASR. For each recognized utterance, the following three lines show the probability of each user intent, denoted by $Pr$. Finally, the last line, denoted by $a$, shows the performed action by SmartWheeler.

For instance, the second utterance shows that the user actually uttered *turn right a little*, but it is recognized as *10 writer little* by ASR. The most probable intent returned by HTMM for this utterance is $i_3$ : *turn right little* with 0.99 probability. This is because HTMM considers Markovian property for deriving intents, cf. Sect. 4.2.1. Consequently, in the second turn the intent $i_3$ gets high probability since in the first turn the user intent is $i_3$ with high probability.

Before we learn a complete dialog POMDP, first we learned a dialog MDP using the SmartWheeler dialogs. We used the learned intents, $i_1, \ldots, i_{11}$, as the states of the MDP. The learned states are presented in Table 6.5. Note that for the intent $i_7$, we used it as the state for the command *turn degree right* as in the intent $i_7$ the word *right* occurs with slightly higher probability than the word *left*.

Then, we learned the transition model, i.e., the smoothed maximum likelihood transition method, introduced in Sect. 4.3. Note that the dialog MDP here is in fact an intent dialog MDP in the same way defined in Sect. 4.4. That is, we used a deterministic intent observation model for the dialog MDP, which considers the observed intent as its current state during the dialog interaction.

**Table 6.4** A sample from the results of applying HTMM on SmartWheeler

| | |
|---|---|
| | ... |
| $u_1$ : | turn right |
| $\widetilde{u}_1$ : | [turn right] |
| $\Pr_1$ | $i_1 : 7.1e-9$ $i_2 : 9.6e-10$ $i_3 : \mathbf{0.6}$<br>$i_4 : 0.2$ $i_5 : 2.6e-8$ $i_6 : 2.2e-5$<br>$i_7 : 0.1$ $i_8 : 6.3e-7$ $i_9 : 1.6e-8$ $i_{10} : 2.4e-6$ $i_{11} : 5.2e-9$ |
| $a_1$ : | TURN RIGHT |
| $u_2$ : | turn right a little |
| $\widetilde{u}_2$ : | [10 writer little] |
| $\Pr_2$ | $i_1 : 0.0$ $i_2 : 0.0$ $i_3 : \mathbf{0.9}$<br>$i_4 : 0.0$ $i_5 : 1.3e-7$ $i_6 : 5.8e-8$<br>$i_7 : 8.8e-8$ $i_8 : 1.2e-6$ $i_9 : 5.9e-5$ $i_{10} : 8.8e-5$ $i_{11} : 1.1e-7$ |
| $a_2$ : | PLEASE REPEAT YOUR COMMAND |
| $u_3$ : | turn right a little |
| $\widetilde{u}_3$ : | [turn right to lead a] |
| $\Pr_3$ | $i_1 : 6.1e-11$ $i_2 : 9.5e-12$ $i_3 : \mathbf{0.9}$<br>$i_4 : 0.0$ $i_5 : 2.7e-08$ $i_6 : 2.0e-07$<br>$i_7 : 0.0$ $i_8 : 3.9e-9$ $i_9 : 1.9e-10$ $i_{10} : 4.4e-08$ $i_{11} : 1.7e-11$ |
| $a_3$ : | TURN RIGHT A LITTLE |
| $u_4$ : | stop |
| $\widetilde{u}_4$ : | [stop] |
| $\Pr_4$ | $i_1 : 3.2e-5$ $i_2 : 4.8e-6$ $i_3 : 0.0$<br>$i_4 : 0.0$ $i_5 : 0.0$ $i_6 : 7.8e-6$<br>$i_7 : 0.0$ $i_8 : 0.0$ $i_9 : 0.0$ $i_{10} : 0.0$ $i_{11} : \mathbf{0.9}$ |
| $a_4$ : | STOP |
| | ... |

### 6.2.1 *Observation Model Learning*

Built off the learned dialog MDP, we developed two dialog POMDPs by learning the two observation sets and their subsequent observation models: keyword model

**Table 6.5** The SmartWheeler learned states

| | |
|---|---|
| $s_1$ | *move-forward-little* |
| $s_2$ | *move-backward-little* |
| $s_3$ | *turn-right-little* |
| $s_4$ | *turn-left-little* |
| $s_5$ | *follow-left-wall* |
| $s_6$ | *follow-right-wall* |
| $s_7$ | *turn-degree-right* |
| $s_8$ | *go-door* |
| $s_9$ | *set-speed* |
| $s_{10}$ | *follow-person* |
| $s_{11}$ | *stop* |

and intent model, proposed in Sect. 4.4. From these models, we then developed the keyword dialog POMDP and the intent dialog POMDP for SmartWheeler. As mentioned in Sect. 4.5.2, here we show the two observation sets for SmartWheeler and then compare the intent POMDP performance to the keyword POMDP performance.

The keyword observation model for each state uses a keyword that best represents the state. We use the *1-top* word of each state, shown in Table 6.3, as observations (the highlighted words). That is, the observations are:

*forward, backward, right, left, turn, go, for, top, stop.*

Note that states $s_3$ and $s_6$ share the same keyword observation, i.e. *right*. Also, states $s_4$ and $s_5$ share the same keyword observation, i.e., *left*.

For the intent model, each state itself is the observation. Then, the set of observations is equivalent to the set of intents. For SmartWheeler the intent observations are:

$$i_1o,\ i_2o,\ i_3o,\ i_4o,\ i_5o,\ i_6o,\ i_7o,\ i_8o,\ i_9o,\ i_{10}o,\ i_{11}o$$

respectively for the states:

$$s_1,\ s_2,\ s_3,\ s_4,\ s_5,\ s_6,\ s_7,\ s_8,\ s_9,\ s_{10},\ s_{11}.$$

Table 6.6 shows the sample dialog from SmartWheeler after learning the two observation sets. In this table, line $o_1$ is the observation for the recognized utterance by ASR, $\tilde{u}_1$. If the keyword observation model is used the observation will be *right*, however, if intent observation model is used then the observation will be the one inside parenthesis, i.e., $i_3o$. In fact, $i_3o$ is an observation with high probability for the state $s_3$, and with low probability for the rest of states.

Note that in $o_2$ for the case of keyword observation, the observation is *confusedObservation*. This is because for the keyword model, none of the keyword

**Table 6.6** A sample from the results of applying the two observation models on the SmartWheeler dialogs

| | |
|---|---|
| | ... |
| $u_1$ : | turn right |
| $\tilde{u}_1$ : | [turn right] |
| $o_1$ : | *right* $(i_3o)$ |
| $u_2$ : | turn right a little |
| $\tilde{u}_2$ : | [10 writer little] |
| $o_2$ : | *confusedObservation* $(i_3o)$ |
| $u_3$ : | turn right a little |
| $\tilde{u}_3$ : | [turn right to lead a] |
| $o_3$ : | *right* $(i_3o)$ |
| $u_4$ : | stop |
| $\tilde{u}_4$ : | [stop] |
| $o_4$ : | *stop* $(i_{11}o)$ |
| | ... |

observations occurs in the recognized utterance $\tilde{u}_2$. However, the intent observation interestingly becomes $i_3o$ which is the same as the intent observation in $o_1$.

### 6.2.2 *Comparison of the Intent POMDP to the Keyword POMDP*

As mentioned in Sect. 4.5.2, we compared the keyword POMDP to the intent POMDP. Recall from the previous section that in the keyword POMDP, the observation set is the set of learned keywords and the observation model is the learned keyword observation model. In the intent POMDP, however, the observation set is the set of learned intents and the observation model is the learned intent observation model. The learned keyword and intent POMDPs are then compared based on their policies. To do so, we assumed a reward function for the two dialog POMDPs and compared the optimal policies of the two POMDPs, based on their accumulated mean rewards in simulation runs.

Similar to the previous work of Png and Pineau (2011), we considered reward of $+1$ for the SmartWheeler performing the right action at each state, and 0 otherwise. Moreover, for the general query, PLEASE REPEAT YOUR COMMAND, the reward is considered as $+0.4$ for each state where this query occurs. The intuition for this reward is that in each state it is best to perform the right action of the state, and it is better to perform a general query action than to perform any other wrong action in the state. That is the reason for defining the $+0.4$ reward for the query action ($0<+0.4<1$). This reward function is represented in Table 6.9 (top), which is also used as the expert reward function in the IRL experiments in Sect. 6.3.

**Table 6.7** The performance of the intent POMDP vs. the keyword POMDP, learned from the SmartWheeler dialogs

| | Mean reward | *Conf95Min* | *Conf95Max* |
|---|---|---|---|
| Intent POMDP | 8.914 | 8.904 | 8.922 |
| Keyword POMDP | 4.784 | 4.767 | 4.802 |

The dialog POMDP models consist of 11 states, 12 actions, and 10 observations if the keyword observation model is used (9 keywords and the *confusedObservation*). Otherwise, there are 11 observations for the intent observation model. We solved our POMDP models, using ZMDP software available online at: http://www.cs.cmu.edu/~trey/zmdp/. We set a uniform distribution on states, and set the discount factor to 0.90.

Similar to Sect. 4.5.2, we evaluated our learned observation models based on accumulated mean rewards. This is because the reward function is the same for the intent POMDP and keyword POMDP. Then, the learned policy of each model can reflect the quality of the learned observation model.

We used the default simulation in ZMDP software which simulates the environment by randomly sampling observations and uses the provided observation and transition models. Note that since the transition model is the same for the intent POMDP and keyword POMDP, the accumulated reward by policy of each model can demonstrate the quality of the observation model.

Table 6.7 shows the comparison of the two models based on 1000 simulation runs. The table shows that the intent POMDP accumulates strongly higher mean reward than the keyword POMDP based on 1000 simulation runs by ZMDP software. In Table 6.7, *Conf95Min* and *Conf95Max* are, respectively, the minimum 95 % confidence and the maximum 95 % confidence of the accumulated mean reward. This means that with approximately 95 % confidence the accumulated mean reward occurs inside the interval formed by *Conf95Min* and *Conf95Max*.

As such, we perform the POMDP-IRL experiments for learning the reward function from SmartWheeler dialogs on the learned intent POMDP. Similarly, we perform the MDP-IRL experiments on the learned intent MDP, i.e., the intent POMDP with the deterministic observation model.

## 6.3 Reward Function Learning for SmartWheeler

In this section, we experiment the MDP-IRL algorithm, introduced in Sect. 5.2 and the POMDP-IRL-BT algorithm, proposed in Sect. 5.3.1. As mentioned in Sect. 5.1, the IRL experiments are designed to verify if the introduced IRL methods are able to learn a reward function for the expert policy, where the expert policy is represented as a (PO)MDP policy. That is, the expert policy is the optimal policy of the (PO)MDP with a known model. Thus, similar to Sect. 5.6, we assumed an expert reward function $R^{\pi_E}$ and used the (PO)MDP model to find the expert policy

$\pi_E$. The learned expert policy was used to sample $B$ expert trajectories to be used in the IRL algorithms.

Based on the experiments in the previous section, we selected the intent MDP/POMDP to be used as the underlying MDP/POMDP framework. The intent POMDP consists of 11 states, 24 actions, 11 intent observations, and the learned transition and observation models. The initial belief, $b_0$, is set to the uniform belief. The intent MDP is similar to the intent POMDP, but the observation model is deterministic.

### 6.3.1 Choice of Features

Recall from the previous chapter that IRL needs features to represent the reward function. We propose *keyword* features for applying IRL on the learned dialog MDP/POMDP from SmartWheeler. The keyword features are SmartWheeler keywords, i.e., 1-top words for each user intent from Table 6.3. There are nine learned keywords:

*forward, backward, right, left, turn, go, for, top, stop.*

The keyword features for each state of SmartWheeler dialog POMDP are represented in a vector, as shown in Table 6.8. The highlighted values show that states $s_3$ (*turn-right-little*) and $s_6$ (*follow-right-wall*) share the same features, i.e., *right*. Moreover, states $s_4$ (*turn-left-little*) and $s_5$ (*follow-left-wall*) share the same feature, i.e., *left*. In our experiments, we used *keyword-action-wise* features. Such features include the indicator functions for each pair of state-keyword and action. Thus, the feature size for SmartWheeler equals $216 = 9 \times 24$ (9 keywords and 24 actions).

Note that the choice of features is application dependent. The reason for using keywords as state features is that in the intent-based dialog applications the states are the dialog intents, where each intent is described as a vector of k-top words from the domain dialogs. Therefore, the keyword features are relevant features for the states.

Note also that although the keyword features are similar to the keyword observations proposed for POMDP observations in Sect. 4.4, there is no explicit learned model for their dynamics such as the keyword observation model proposed in Sect. 4.4. In particular, for MDPs there is no observation model, however the keyword features are used in MDP-IRL for the reward function representation.

**Table 6.8** Keyword features for the SmartWheeler dialogs. The highlighted values show that states $s_3$ (turn-right-little) and $s_6$ (follow-right-wall) share the same keyword (right), and states $s_4$ (turn-left-little) and $s_5$ (follow-left-wall) share the same keyword (left)

| | forward | backward | **right** | **left** | turn | go | for | top | stop |
|---|---|---|---|---|---|---|---|---|---|
| $s_1$ | 1 | 0 | 0 | 0 | 0 | 0 | 0 | 0 | 0 |
| $s_2$ | 0 | 1 | 0 | 0 | 0 | 0 | 0 | 0 | 0 |
| $s_3$ | 0 | 0 | **1** | 0 | 0 | 0 | 0 | 0 | 0 |
| $s_4$ | 0 | 0 | 0 | **1** | 0 | 0 | 0 | 0 | 0 |
| $s_5$ | 0 | 0 | 0 | **1** | 0 | 0 | 0 | 0 | 0 |
| $s_6$ | 0 | 0 | **1** | 0 | 0 | 0 | 0 | 0 | 0 |
| $s_7$ | 0 | 0 | 0 | 0 | 1 | 0 | 0 | 0 | 0 |
| $s_8$ | 0 | 0 | 0 | 0 | 0 | 1 | 0 | 0 | 0 |
| $s_9$ | 0 | 0 | 0 | 0 | 0 | 0 | 1 | 0 | 0 |
| $s_{10}$ | 0 | 0 | 0 | 0 | 0 | 0 | 0 | 1 | 0 |
| $s_{11}$ | 0 | 0 | 0 | 0 | 0 | 0 | 0 | 0 | 1 |

## 6.3.2 MDP-IRL Learned Rewards

In this section, we show the learned reward function by the MDP-IRL algorithm for the expert policy, where similar to previous works (Ng and Russell 2000; Choi and Kim 2011), the expert policy is an MDP policy (cf. Sect. 5.1). To do so, we assumed an expert reward function for the learned intent MDP from SmartWheeler. We then solved the model to find the (near) optimal policy which is used as the expert policy.

Similar to the previous section, we assumed the reward function used in Png and Pineau (2011). Table 6.9 (top) shows the expert reward function. That is, we considered +1 reward for performing the right action at each state, and 0 otherwise. Moreover, for the general query PLEASE REPEAT YOUR COMMAND in every state the reward is considered as +0.4. We then solved the intent MDP model with the assumed expert reward to find the optimal policy, i.e., the expert policy. The expert policy for each of the MDP state is represented in Table 6.10. Interestingly, the expert policy suggests performing the right action of each state.

We then applied the MDP-IRL algorithm on SmartWheeler dialog MDP described above using the introduced keyword features in Table 6.8. The algorithm was able to learn a reward function in which the policy equals the expert policy for all states (the expert policy shown in Table 6.10). Table 6.9 (bottom) shows the learned reward function. Comparing the assumed expert reward function in Table 6.9 (top) to the learned reward function in Table 6.9 (bottom), we observe that the rewards in the two tables are different. However, the policy for the learned reward is exactly the same as the policy for the assumed reward (shown in Table 6.10). Note that the highlighted values show that the reward of performing action $a_3$ (TURNT RIGHT A LITTLE) and action $a_6$ (FOLLOW THE RIGHT WALL) is the same for state $s_3$ (turn-right-little). Similarly, the highlighted values show that the reward of performing action $a_3$ (TURNT RIGHT A LITTLE) and action $a_6$ (FOLLOW THE RIGHT WALL) is the same for state $s_6$ (follow-right-wall). This is probably because states $s_3$ and $s_6$ share the same feature, i.e., the same keyword (right). However, the policy for the learned reward suggests performing action $a_3$ (TURNT RIGHT A LITTLE) in state $s_3$ (turn-right-little) and action $a_6$ (FOLLOW

**Table 6.9** *Top*: The assumed expert reward function for the dialog MDP/POMDP learned from SmartWheeler dialogs. *Bottom*: The learned reward function for the learned dialog MDP from SmartWheeler dialogs using keyword features. The highlighted values show that the reward of performing action $a_3$ (TURNT RIGHT A LITTLE) and action $a_6$ (FOLLOW THE RIGHT WALL) is the same for state $s_3$ (turn-right-little). Similarly, the highlighted values show that the reward of performing action $a_3$ (TURNT RIGHT A LITTLE) and action $a_6$ (FOLLOW THE RIGHT WALL) is the same for state $s_6$ (follow-right-wall). However, the policy for the learned reward suggests performing action $a_3$ (TURNT RIGHT A LITTLE) in state $s_3$ (turn-right-little) and action $a_6$ (FOLLOW THE RIGHT WALL) in state $s_6$ (follow-right-wall). Specifically, the policy for the learned reward is exactly the same as the policy for the assumed reward (shown in Table 6.10). There is a similar pattern for states $s_4$ (turn-left-little) and $s_5$ (follow-left-wall) and actions $a_4$ (TURNT LEFT A LITTLE) and $a_5$ (FOLLOW THE LEFT WALL)

| Assumed expert reward function | | | | | | | | | | | | | |
|---|---|---|---|---|---|---|---|---|---|---|---|---|---|
| | $a_1$ | $a_2$ | $a_3$ | $a_4$ | $a_5$ | $a_6$ | $a_7$ | $a_8$ | $a_9$ | $a_{10}$ | $a_{11}$ | $a_{12}$ | ... | REPEAT |
| $s_1$ | 1.0 | 0 | 0 | 0 | 0 | 0 | 0 | 0 | 0 | 0 | 0 | 0 | ... | 0.4 |
| $s_2$ | 0 | 1.0 | 0 | 0 | 0 | 0 | 0 | 0 | 0 | 0 | 0 | 0 | ... | 0.4 |
| $s_3$ | 0 | 0 | 1.0 | 0 | 0 | 0 | 0 | 0 | 0 | 0 | 0 | 0 | ... | 0.4 |
| $s_4$ | 0 | 0 | 0 | 1.0 | 0 | 0 | 0 | 0 | 0 | 0 | 0 | 0 | ... | 0.4 |
| $s_5$ | 0 | 0 | 0 | 0 | 1.0 | 0 | 0 | 0 | 0 | 0 | 0 | 0 | ... | 0.4 |
| $s_6$ | 0 | 0 | 0 | 0 | 0 | 1.0 | 0 | 0 | 0 | 0 | 0 | 0 | ... | 0.4 |
| $s_7$ | 0 | 0 | 0 | 0 | 0 | 0 | 1.0 | 0 | 0 | 0 | 0 | 0 | ... | 0.4 |
| $s_8$ | 0 | 0 | 0 | 0 | 0 | 0 | 0 | 1.0 | 0 | 0 | 0 | 0 | ... | 0.4 |
| $s_9$ | 0 | 0 | 0 | 0 | 0 | 0 | 0 | 0 | 1.0 | 0 | 0 | 0 | ... | 0.4 |
| $s_{10}$ | 0 | 0 | 0 | 0 | 0 | 0 | 0 | 0 | 0 | 1.0 | 0 | 0 | ... | 0.4 |
| $s_{11}$ | 0 | 0 | 0 | 0 | 0 | 0 | 0 | 0 | 0 | 0 | 1.0 | 0 | ... | 0.4 |
| Learned reward function by MDP-IRL | | | | | | | | | | | | | | |
| $s_1$ | 1.0 | 0 | 0 | 0 | 0 | 0 | 0 | 0 | 0 | 0 | 0 | 0 | ... | 0 |
| $s_2$ | 0 | 1.0 | 0 | 0 | 0 | 0 | 0 | 0 | 0 | 0 | 0 | 0 | ... | 0 |
| $s_3$ | 0 | 0 | **1.0** | 0 | 0 | **1.0** | 0 | 0 | 0 | 0 | 0 | 0 | ... | 0 |
| $s_4$ | 0 | 0 | 0 | **1.0** | **1.0** | 0 | 0 | 0 | 0 | 0 | 0 | 0 | ... | 0 |
| $s_5$ | 0 | 0 | 0 | **1.0** | **1.0** | 0 | 0 | 0 | 0 | 0 | 0 | 0 | ... | 0 |
| $s_6$ | 0 | 0 | **1.0** | 0 | 0 | **1.0** | 0 | 0 | 0 | 0 | 0 | 0 | ... | 0 |
| $s_7$ | 0 | 0 | 0 | 0 | 0 | 0 | 1.0 | 0 | 0 | 0 | 0 | 0 | ... | 0 |
| $s_8$ | 0 | 0 | 0 | 0 | 0 | 0 | 0 | 1.0 | 0 | 0 | 0 | 0 | ... | 0 |
| $s_9$ | 0 | 0 | 0 | 0 | 0 | 0 | 0 | 0 | 1.0 | 0 | 0 | 0 | ... | 0 |
| $s_{10}$ | 0 | 0 | 0 | 0 | 0 | 0 | 0 | 0 | 0 | 1.0 | 0 | 0 | ... | 0 |
| $s_{11}$ | 0 | 0 | 0 | 0 | 0 | 0 | 0 | 0 | 0 | 0 | 1.0 | 0 | ... | 0 |

THE RIGHT WALL) in state $s_6$ (follow-right-wall). There is a similar pattern for states $s_4$ (turn-left-little) and $s_5$ (follow-left-wall) and actions $a_4$ (TURNT LEFT A LITTLE) and $a_5$ (FOLLOW THE LEFT WALL). However, the policy of the learned reward function is exactly the same as expert policy, i.e. the policy for the assumed reward (shown in Table 6.10). Therefore, though the learned reward function is different from the assumed reward function, the policy of the learned reward is the

**Table 6.10** The policy of the learned dialog MDP from SmartWheeler dialogs with the assumed expert reward function

| State | State description | Expert action | Expert action description |
|---|---|---|---|
| $s_1$ | *move-forward-little* | $a_1$ | DRIVE FORWARD A LITTLE |
| $s_2$ | *move-backward-little* | $a_2$ | DRIVE BACKWARD A LITTLE |
| $s_3$ | *turn-right-little* | $a_3$ | TURN RIGHT A LITTLE |
| $s_4$ | *turn-left-little* | $a_4$ | TURN LEFT A LITTLE |
| $s_5$ | *follow-left-wall* | $a_5$ | FOLLOW THE LEFT WALL |
| $s_6$ | *follow-right-wall* | $a_6$ | FOLLOW THE RIGHT WALL |
| $s_7$ | *turn-degree-right* | $a_7$ | TURN RIGHT DEGREES |
| $s_8$ | *go-door* | $a_8$ | GO THROUGH THE DOOR |
| $s_9$ | *set-speed* | $a_9$ | SET SPEED TO MEDIUM |
| $s_{10}$ | *follow-wall* | $a_{10}$ | FOLLOW THE WALL |
| $s_{11}$ | *stop* | $a_{11}$ | STOP |

same as the policy of the assumed reward for expert policy. Recall that IRL is an ill-posed problem, as mentioned in Sect. 5.1. There could be many reward functions that makes expert policy optimal.

### 6.3.3 POMDP-IRL-BT Evaluation

In this section, we show our experiments on the POMDP-IRL-BT algorithm on the intent dialog POMDP learned from SmartWheeler. As mentioned earlier, to evaluate the IRL algorithms, we consider that expert policy is a POMDP policy using an assumed reward function. Similar to previous section, we assumed that the expert reward function is the one represented in Table 6.9 (top). For the choice of features, we also used the keyword features shown in Table 6.8.

Similar to the experiments in Sect. 5.6, we performed two fold cross validation experiments by generating 10 expert trajectories. The expert trajectories are truncated after 20 steps, since there is no terminal state here. We then used the Perseus software with the same setting as described in Sect. 5.6. That is, we set the solver to use 10,000 random samples for solving the optimal policy of each candidate reward. The other parameter is max-time for execution of the algorithm, which is set to 1000.

Based on the specification above, we performed POMDP-IRL-BT on SmartWheeler expert trajectory for training. The experimental results showed that the policy of the learned reward was the same as the expert policy for 194 beliefs inside the testing trajectory out of the 200 beliefs, i.e., 97 % matched actions. For all the 6 errors, the expert action was TURN RIGHT LITTLE, i.e., the right action for the state TURN-RIGHT-LITTLE, while the action of the learned reward suggested FOLLOW RIGHT WALL. However, this error did not happen in all the cases which the expert action was TURN RIGHT LITTLE in the testing trajectory.

Afterwards, we used state-action-wise features as defined in Sect. 5.6. Such features include an indicator function for each state-action pair. In SmartWheeler, there are 11 states and 24 actions, then the size of state-action-wise features equals $264 = 11 \times 24$. This is a slight increase compared to the size of keyword features, i.e., 216. We observed that in our experiment the learned policy is exactly the same as the expert policy for the 200 beliefs inside the testing trajectory using state-action-wise features, i.e., 100 % matched with the expert policy. In words, POMDP-IRL-BT was able to learn a reward function for the expert policy using the learned dialog POMDP from SmartWheeler dialogs. In the following section, we compare POMDP-IRL-BT to POMDP-IRL-MC introduced in Sect. 5.5, in which the policy values are estimated using the Monte Carlo estimator rather than by approximating the belief transitions.

### 6.3.4 Comparison of POMDP-IRL-BT to POMDP-IRL-MC

In Sect. 5.4, we saw that Choi and Kim (2011) proposed IRL algorithms in POMDP framework by assuming policies in the form of an FSC and thus using PBPI (point-based policy iteration) (Ji et al. 2007), as POMDP solver. In their algorithm, they used Monte Carlo estimator to estimate the value of expert policy whereas we used an estimated belief transition model for the expert beliefs to be able to use Bellman equation for approximating the expert policy values as well as candidate policy values. As stated in Sect. 5.5, we also implemented the Monte Carlo estimator (Eq. (5.19)) for the estimation of policy values in Line 7 in Algorithm 8, and used the Perseus software (Spaan and Vlassis 2005) as the POMDP solver. This new algorithm is called POMDP-IRL-MC. We compared POMDP-IRL-BT to POMDP-IRL-MC. The purpose of such experiments was to compare the belief transition estimation to the Monte Carlo estimation.

We compared the two algorithms, POMDP-IRL-BT and POMDP-IRL-MC, based on the following criteria:

1. Percentage of the learned actions that matches to the expert actions.
2. Value of learned policy with respect to the value of expert policy.
3. CPU time spent by the algorithm as the number of expert trajectories (training data) increases.

Criteria 1 and 2 are used to evaluate the quality of the learned reward function for the expert. As in the previous experiment, the higher the matched actions, the better the learned reward function is. Similarly, criterion 2 compares the value of the learned reward function with the value of expert reward function. The higher the value of the learned policy, the better the learned reward function is. The results for these criteria is based on two fold cross validation using 400 expert trajectories, i.e., each fold contains of 200 expert trajectories.

Note that the value of learned policy (in criterion 2) is the sampled value of the policy. This was done by running the policy starting from a uniform belief to the maximum $maxT = 20$ time step or until it reaches the terminal state. The sampled values are averaged over 100 runs, and are calculated using:

$$\hat{V}^{\pi}(b) = \left[ \sum_{t=0}^{maxT} \gamma^t R(b_t, \pi(b_t)) | \pi, b_0 = b \right].$$

Finally, criterion 3 evaluates the CPU time spent by the algorithm as the number of expert trajectories increases. This is to verify which of the two algorithms, POMDP-IRL-BT and POMDP-IRL-MC, requires more computation time. Below, we report on our experiments on SmartWheeler domain based on the above-mentioned criteria.

#### 6.3.4.1 Evaluation of the Quality of the Learned Rewards

First, we evaluated POMDP-IRL-BT and POMDP-IRL-MC using keyword features based on criteria 1 and 2. The results are shown in Fig. 6.2 (top) and (bottom). The two figures show consistent results in which the performance of POMDP-IRL-BT and POMDP-IRL-MC is comparable.

Figure 6.2 (top) shows percentage of the matched actions to those of expert, as the number of iterations increases (the first criteria). The figure demonstrates that after around 15 iterations the learned actions for 95 % of testing trajectories matches to actions suggested by the expert policy, in both the POMDP-IRL-BT and POMDP-IRL-MC algorithms. The figure also shows that after iteration 15, percentage of the matched actions fluctuates slightly as the number of iterations increases, however percentage remains above 90 %.

Moreover, Fig. 6.2 (bottom) plots the value of the learned policy (the sampled value) as the number of iterations increases (criterion 2). Similar to Fig. 6.2 (top), we observe that for both POMDP-IRL-BT and POMDP-IRL-MC after iteration 15 the learned policy value becomes close to the expert policy value. Moreover, though the learned policy values fluctuate slightly, it remains close to the expert policy value after iteration 15.

The reason for these fluctuations is the choice of features. In the experiments reported above we used the automatically learned keyword features for our POMDP-IRL experiments. In Table 6.8, we saw that the states 3 and 6 share the same feature *right*. Similarly, the states 4 and 5 share the same feature *left*. Although this kind of feature sharing can reduce the size of features, it can lead to learning wrong actions for the sharing states.

Therefore, we performed similar experiments on SmartWheeler but this time using state-action features. These features include the indicator functions for each pair of state and action. Thus, the feature size for SmartWheeler equals $11 \times 24 = 264$, which is a slight increase compared to the size of keyword features, i.e., 216. Similar to the keyword features, we evaluated state-action features on SmartWheeler based on criteria 1 and 2. The results are shown in Fig. 6.3 (top) and (bottom).

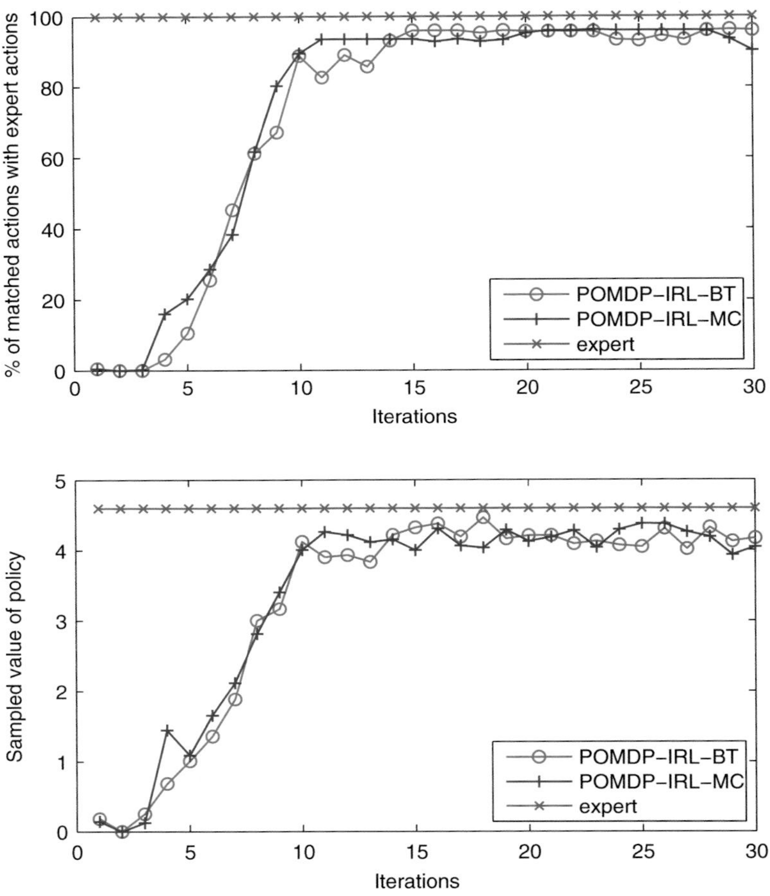


**Fig. 6.2** Comparison of the POMDP-IRL algorithms using keyword features on the learned dialog POMDP from SmartWheeler. *Top*: percentage of matched actions. *Bottom*: sampled value of the learned policy

Figure 6.3 (top) and (bottom) show consistent results in which the performance of POMDP-IRL-BT reaches to expert performance. Figure 6.3 (top) shows percentage of the matched actions between the learned and expert policies, as the number of iterations increases. The figure shows that this percentage reaches to 100 % in POMDP-IRL-BT, while it reaches to 97 % in POMDP-IRL-MC.

Moreover, Fig. 6.3 (bottom) plots the value of the learned policy as the number of iterations increases. We observe that the learned value equals the value of expert policy in POMDP-IRL-BT (at iteration 13), while in POMDP-IRL-MC it only gets close to the value of expert policy (at iteration 17). Furthermore, Fig. 6.3 (top) and (bottom) show that using state-action features, POMDP-IRL-BT reaches its optimal performance (equal to the expert performance) slightly earlier than POMDP-IRL-MC (at iteration 13 and iteration 17, respectively).

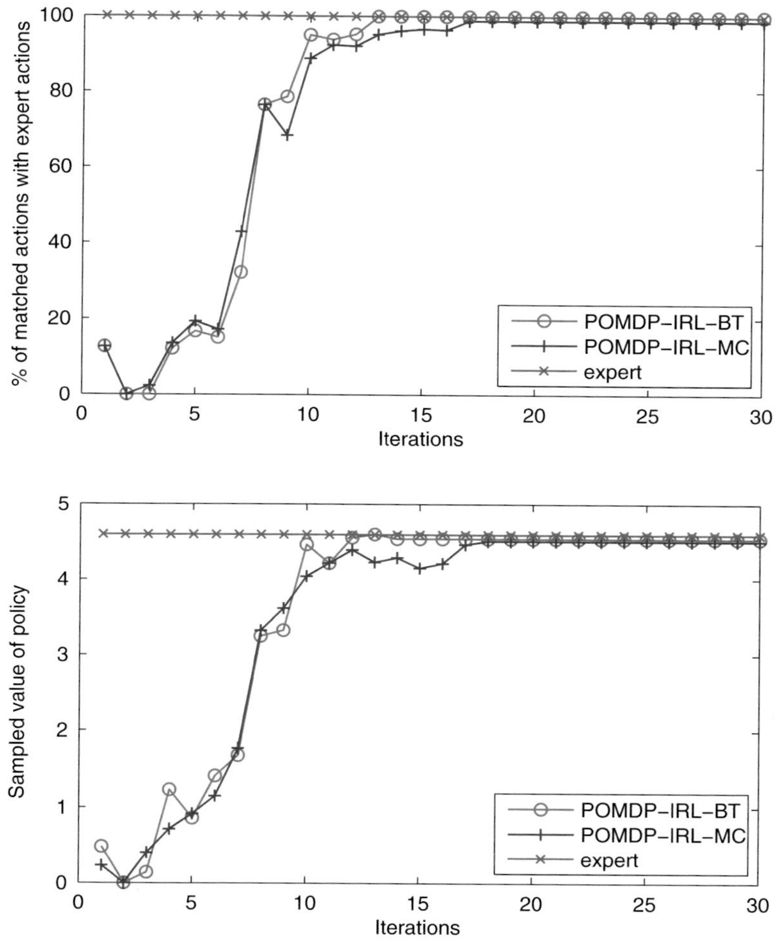


**Fig. 6.3** Comparison of the POMDP-IRL algorithms using state-action-wise features on the learned dialog POMDP from SmartWheeler. *Top*: percentage of matched actions. *Bottom*: sampled value of learned policy

#### 6.3.4.2 Evaluation of the Spent CPU Time

Figure 6.4 demonstrates the spent time by POMDP-IRL-BT and POMDP-IRL-MC as the number of expert trajectories (training data) increases. The results show that by increasing the number of expert trajectories, POMDP-IRL-BT requires considerably more time than POMDP-IRL-MC. Note that the figure plots the spent time by the number of trajectories in the log base. This increase is due to increase of the size of belief transition matrix, Eq. (5.12), as the number of expert trajectories increases. In other words, the belief transition matrix requires much more time to

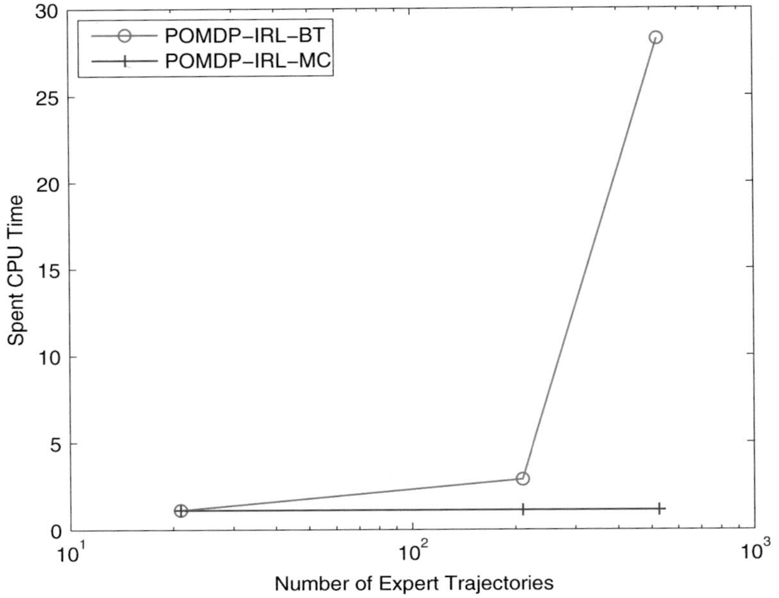


**Fig. 6.4** Spent CPU time by POMDP-IRL algorithms on SmartWheeler, as the number of expert trajectories (training data) increases

be constructed as the number of beliefs in expert trajectories increases. Also, note that this matrix is constructed for each candidate policy, which in turn increases the CPU time.

In sum, our experimental results showed that using state-action features, the POMDP-IRL-BT is able to learn a reward function in which the policy matches the expert policy for 100 % of beliefs in the testing trajectories, while POMDP-IRL-MC learned a reward function in which the policy matched the expert policy for only 97 % of beliefs in testing trajectories. However, POMDP-IRL-MC *does* scale substantially better than POMDP-IRL-BT. In the case of large number of expert trajectories, POMDP-IRL-BT can still be useful. For instance, we can use all expert trajectories to estimate the transition and observation models, but, select part of the expert trajectories to learn the reward function.

## 6.4 Conclusions

In this chapter, we applied the proposed methods in this book on a healthcare dialog management. We used the dialogs collected by an intelligent wheelchair called SmartWheeler for learning the model components of the dialog POMDP. To do so, we first learned the user intents that occurred in the SmartWheeler dialogs and used them as states of the dialog POMDP. Then, we used the learned states and the extracted SmartWheeler actions to learn the maximum likelihood transition model. For the observation model of SmartWheeler dialog POMDP, we learned both the intent and keyword observation models. We observed that the intent POMDP, i.e., the POMDP using the intent observation model, performed significantly better than the keyword POMDP.

We then introduced the automatically learned keyword features and applied the MDP-IRL algorithm, introduced in the previous chapter, on the learned intent MDP from SmartWheeler. The algorithm learned a reward function whose policy completely matched to the expert policy using the keyword-action-wise features. Furthermore, we evaluated our proposed POMDP-IRL-BT algorithm on the learned intent POMDP from SmartWheeler. We observed that POMDP-IRL-BT is able to learn a reward function that accounts for the expert policy using keyword-action-wise and state-action-wise features.

Finally, we compared the POMDP-IRL-BT algorithm to the POMDP-IRL-MC algorithm which uses Monte Carlo estimation in the place of belief transition estimation. Our experiments showed that the both algorithms are able to learn a reward function that accounts for the expert policy using keyword-action-wise and state-action-wise features. Furthermore, our experimental results showed that POMDP-IRL-BT slightly outperforms the POMDP-IRL-MC algorithm, however, the POMDP-IRL-MC does scale better than POMDP-IRL-BT.

Overall, the experiments on SmartWheeler dialogs showed that the proposed methods are able to learn the dialog POMDP model components from real dialogs. In the following section, we summarize the book and address multiple avenues for future research of dialog POMDP model learning.

# Chapter 7
# Conclusions and Future Work

## 7.1 Summary

Spoken dialog systems (SDSs) are the systems that help the human user to accomplish a task using the spoken language. Dialog management is a difficult problem since automatic speech recognition (ASR) and natural language understanding (NLU) make errors which are the sources of uncertainty in SDSs. Moreover, the human user behavior is not completely predictable. The users may change their *intents* during the dialog, which makes the SDS environment stochastic. Furthermore, the users may express an intent in several ways which makes dialog management more challenging.

In this context, partially observable Markov decision process (POMDP) framework has been used to model the dialog management of SDSs. The POMDP framework can deal with both the uncertainty and stochasticity in the environment in a principled way. Furthermore, the POMDP framework has shown better performance compared to other frameworks, such as Markov decision processes (MDPs). This is particularly the case in the noisy environments, which is often the case in spoken dialog systems.

However, POMDPs and their application on SDSs involve many challenges. In particular, we were mostly interested in learning the dialog POMDP model components from unannotated and noisy dialogs. In this context, there are a large number of unannotated dialogs available which can be used for learning dialog POMDP model components. In addition, learning the dialog POMDP model components from data is particularly significant since the learned dialog POMDP model directly affects the POMDP policy. Furthermore, learning proper dialog POMDP model components from real data could be highly beneficial since there



H. Chinaei, B. Chaib-draa, *Building Dialogue POMDPs from Expert Dialogues*, SpringerBriefs in Speech technology, DOI 10.1007/978-3-319-26200-0_7

is a rich literature on model-based POMDP solving that can be used once the dialog POMDP model components are learned. In words, if we are able to learn a realistic dialog POMDP from data, then we can make use of available POMDP solvers for learning the POMDP policy.

In this book, we proposed methods for learning dialog POMDP model components from unannotated dialogs for intent-based dialog domains in which the user intent is the dialog state. We demonstrated the big picture of our approach in a descriptive algorithm (Algorithm 1). Our POMDP model learning approach started by learning the dialog POMDP states. The learned states were then used for learning the transition model followed by the dialog POMDP observations and observation model. Building off these learned dialog POMDP model components, we proposed two POMDP-IRL algorithms for learning the reward function.

For the dialog states, we learned the possible user intents that appeared in the user dialogs using an unsupervised topic modeling method. In this way, we were able to learn the user intents from unannotated dialogs and used them as the dialog POMDP states. To do so, we used HTMM (hidden topic Markov model) which is a variation of latent Dirichlet allocation (LDA) that considers the Markovian property between dialogs. Using the learned intents as the dialog states, and the set of actions, extracted from the dialogs, we learned a maximum likelihood transition model for the dialog POMDP. We then proposed two observation models: the keyword model and the intent model. The keyword model used only the learned keywords, from the topic modeling approach, as the set of observations. The intent model, however, used the set of intents as the set of observations. As the two models include a small number of observations, solving the POMDP model becomes tractable.

Furthermore, we introduced trajectory-based inverse reinforcement learning (IRL) for learning the reward function in the (PO)MDP framework using expert trajectories. In this context, we introduced the MDP-IRL algorithm, the basic IRL algorithm in the MDP framework. We then proposed two POMDP-IRL algorithms: POMDP-IRL-BT and PB-POMDP-IRL. The POMDP-IRL-BT algorithm is similar to the MDP-IRL. However, POMDP-IRL-BT uses belief states rather states, and approximates a belief transition model, which is similar to the state transition model in MDPs. On the other hand, PB-POMDP-IRL is a point-based POMDP-IRL algorithm that approximates the value of the new beliefs, which occurs in the computation of the policy values, using a linear approximation of expert beliefs. The two algorithms are able to learn a reward function that accounts for expert policy. However, our experimental results showed that POMDP-IRL-BT outperforms PB-POMDP-IRL since the policy of learned reward function by the former algorithm matched with more expert actions.

We then applied the proposed methods in this book to learn a dialog POMDP from dialogs collected in a healthcare domain. That is, we used the dialogs collected by SmartWheeler, an intelligent wheelchair for handicapped people. We were able to learn 11 user intents, which were considered as states of the dialog POMDP. Based on the learned intents and the SmartWheeler actions, we then learned the maximum likelihood transition model. We then learned the two observation sets and their subsequent observation models: the keyword and intent models.

Our experimental results showed that the intent model outperforms the keyword model-based on accumulated mean rewards in simulation runs. We thus used the learned intent POMDP for the rest of experiments, i.e., for IRL evaluations.

To perform the IRL experiments, we introduced the automatically learned keyword features. We then applied the MDP-IRL algorithm, on the learned intent MDP from SmartWheeler. The algorithm learned a reward function whose policy completely matched to the expert policy using the keyword-action-wise features. Furthermore, we evaluated the POMDP-IRL-BT algorithm on the learned intent POMDP from SmartWheeler. We observed that POMDP-IRL-BT is able to learn a reward function that accounts for the expert policy using keyword-action-wise features.

Finally, we compared the POMDP-IRL-BT algorithm that uses belief transition estimation to the POMDP-IRL-MC algorithm that uses Monte Carlo estimation. Our experimental results showed that both algorithms are able to learn a reward function that accounts for the expert policy. Furthermore, the results showed that POMDP-IRL-BT slightly outperforms the POMDP-IRL-MC algorithm based on matched actions to the expert actions as well as the learned policy values. On the other hand, the POMDP-IRL-MC algorithm does scale better than the POMDP-IRL-BT algorithm.

## 7.2 Future Work

The methods presented in this book can be extended in several directions. In particular, we used HTMM to learn the dialog POMDP intents, mainly because HTMM considers the Markovian property inside dialogs and it is computationally efficient. One direction for future work can be application of other topic modeling approaches such as the LDA (Blei et al. 2003). A survey of topic modeling methods can be found in Blei (2011) and Daud et al. (2010). Moreover, for the transition model we used the add-one smoothed transition model due to its simplicity and sufficiency for the purpose of our experiments. However, there are many other smoothing approaches in the literature which can be tested and compared to the introduced add-one smoothed transition model. For a comprehensive background on smoothing techniques the reader is refereed to Manning and Schütze (1999) and Jurafsky and Martin (2009).

We proposed two sets of observations and their subsequent observation models. The proposed learned observation models could be further extended and enhanced, for instance, by merging the keyword observations and intent observations, considering multiple top keywords of each state rather than considering only one keyword. Furthermore, other methods could be used for learning the observation model such as Bayesian-based methods (Atrash and Pineau 2010; Doshi and Roy 2008; Png and Pineau 2011; Png et al. 2012). In particular, Png and Pineau (2011) and Png et al. (2012) proposed an online Bayesian approach for updating the observation model which can be extended for learning the observation model of dialog POMDPs from SmartWheeler dialogs.

We introduced the basic MDP-IRL algorithm of Ng and Russell (2000), and extended it for POMDPs. However, there are a vast number of IRL algorithms in the MDP framework (Abbeel and Ng 2004; Ramachandran and Amir 2007; Neu and Szepesvári 2007; Syed and Schapire 2008; Ziebart et al. 2008; Boularias et al. 2011). The MDP-IRL algorithms can potentially be extended to POMDPs (Kim et al. 2011). In particular, Kim et al. (2011) extended the MDP-IRL algorithm of Abbeel and Ng (2004), which is called max-margin between feature expectations (MMFE), to a finite state controller (FSC) based POMDP-IRL algorithm. The authors showed that the extension of MMFE for POMDPs performs pretty well based on experiments on several POMDP benchmarks. The MMFE POMDP algorithm of Kim et al. (2011) also could be extended as a point-based POMDP-IRL algorithm in order to take advantage of the computational efficiency of point-based POMDP solvers such as Perseus.

Furthermore, the IRL algorithms require (dialog) features for representing the reward function. A relevant reward function to the dialog system and users can be only learned by studying and extracting relevant features from the dialog domain. Future research should be devoted to automatic methods for learning the relevant and proper features that are suitable for reward representation and reward function learning. We also observed that POMDP-IRL-BT algorithm does not scale as the number of trajectories increases. Although the scalability may not be a great issue as the algorithm can learn the reward function of the expert using a small number of trajectories, another future avenue of research can be enhancing the scalability of the POMDP-IRL-BT algorithm.

Ultimately, in this book, we considered intent-based dialog POMDPs particularly because they can have significant applications, for instance in spoken web search. Our dialog POMDPs currently deal with small set of intents; they can, however, be extended to larger domains, for instance, by considering the domain's hierarchy, and considering a dialog POMDP for each level of the hierarchy. Furthermore, the developed techniques in other dialog domains can be incorporated for intent-based dialog POMDPs, such as factored-based transition and observation model (Williams 2006).

# References


Abbeel, P., & Ng, A. Y. (2004). Apprenticeship learning via inverse reinforcement learning. In *Proceedings of the 21st International Conference on Machine Learning (ICML'04)*, Banff, AB.

Ai, H., & Litman, D. J. (2007). Knowledge consistent user simulations for dialog systems. In *Proceedings of the 8th Annual Conference of the International Speech Communication Association (INTERSPEECH'07)*, Antwerp.

Atrash, A., & Pineau, J. (2010). A Bayesian method for learning POMDP observation parameters for robot interaction management systems. In *The POMDP Practitioners Workshop*.

Balakrishnan, N., & Nevzorov, V. (2004). *A primer on statistical distributions*. John Wiley & Sons.

Bellman, R. (1957a). *Dynamic programming*. Princeton: Princeton University Press.

Bellman, R. (1957b). A Markovian decision process. *Journal of Mathematics and Mechanics, 6*(6), 679–684

Bishop, C. M. (2006). *Pattern recognition and machine learning*. Secaucus, New York: Springer.

Blei, D. (2012). Probabilistic topic models. *Communications of the ACM, 55*(4), 77–84.

Blei, D. M., Ng, A. Y., & Jordan, M. I. (2003). Latent Dirichlet allocation. *Journal of Machine Learning Research, 3*, 993–1022.

Bonet, B., & Geffner, H. (2003). Faster heuristic search algorithms for planning with uncertainty and full feedback. In *Proceedings of the 18th International Joint Conference on Artificial Intelligence (IJCAI'03)*, Acapulco, Mexico.

Boularias, A., Chinaei, H. R., & Chaib-draa, B. (2010). Learning the reward model of dialogue POMDPs from data. In *NIPS 2010 Workshop on Machine Learning for Assistive Technologies*, Vancouver, BC.

Boularias, A., Kober, J., & Peters, J. (2011). Relative entropy inverse reinforcement learning. *Journal of Machine Learning Research - Proceedings Track, 15*, 182–189.

Brown, L. D. (1986). *Fundamentals of statistical exponential families: With applications in statistical decision theory*. Hayworth, CA: Institute of Mathematical Statistics.

Cassandra, A., Kaelbling, L., & Littman, M. (1995). Acting optimally in partially observable stochastic domains. In *Proceedings of the 12th National Conference on Artificial Intelligence (AAAI'95)*, Seattle, Washington.

Chandramohan, S., Geist, M., Lefevre, F., & Pietquin, O. (2011). User simulation in dialogue systems using inverse reinforcement learning. In *Proceedings of the 12th Annual Conference of the International Speech Communication Association (INTERSPEECH'11)*, Florence.

Chinaei, H. (2010). Inverse reinforcement learning for dialogue management. http://www.damas.ift.ulaval.ca/_seminar/filesA10/DAMAS_Seminar_2010-09-16_slides.pdf




H. Chinaei, B. Chaib-draa, *Building Dialogue POMDPs from Expert Dialogues*,
SpringerBriefs in Speech technology, DOI 10.1007/978-3-319-26200-0

Chinaei, H. (2013). *Learning dialogue POMDP model components from expert dialogues*. Ph.D. thesis, Computer Science and Software Engineering Department, Université Laval.

Chinaei, H., & Chaib-Draa, B. (2014a). Dialogue POMDP components (part ii): Learning the reward function. *International Journal of Speech Technology, 17*(4), 325–340.

Chinaei, H. R., & Chaib-Draa, B. (2011). Learning dialogue POMDP models from data. In *Proceedings of the 24th Canadian Conference on Advances in Artificial Intelligence (Canadian AI'11)*, St. John's, Newfoundland.

Chinaei, H. R., & Chaib-Draa, B. (2012). An inverse reinforcement learning algorithm for partially observable domains with application on healthcare dialogue management. In *11th International Conference on Machine Learning and Applications (ICMLA'2012)*, Boca Raton, FL.

Chinaei, H. R., & Chaib-Draa, B. (2014b). Dialogue POMDP components (part i): Learning states and observations. *International Journal of Speech Technology, 17*(4), 309–323.

Chinaei, H. R., Chaib-Draa, B., & Chaib, B. (2014). Dialogue strategy learning in healthcare: A systematic approach for learning dialogue models from data. In *Proceedings of the 5th Workshop on Speech and Language Processing for Assistive Technologies (SLPAT'14), Association for Computational Linguistics (ACL)*, Baltimore, MD.

Chinaei, H. R., Chaib-Draa, B., & Lamontagne, L. (2009). Learning user intentions in spoken dialogue systems. In *Proceedings of the 1st International Conference on Agents and Artificial Intelligence (ICAART'09)*, Porto.

Chinaei, H. R., Chaib-draa, B., & Lamontagne, L. (2012). Learning observation models for dialogue POMDPs. In *Proceedings of the 24th Canadian Conference on Advances in Artificial Intelligence (Canadian AI'12)*, Toronto.

Choi, J., & Kim, K.-E. (2011). Inverse reinforcement learning in partially observable environments. *Journal of Machine Learning Research, 12*, 691–730.

Church, K. W. (1988). A stochastic parts program and noun phrase parser for unrestricted text. In *Proceedings of the 2nd Conference on Applied Natural Language Processing (ANLP'88)*, Austin, TX.

Clark, H., & Brennan, S. (1991). Grounding in communication. *Perspectives on Socially Shared Cognition, 13*(1991), 127–149.

Cuayáhuitl, H., Renals, S., Lemon, O., & Shimodaira, H. (2005). Human-computer dialogue simulation using hidden Markov models. In *Proceedings of IEEE Workshop on Automatic Speech Recognition and Understanding (ASRU'05)*, San Juan, PR.

Cuayáhuitl, H., van Otterlo, M., Dethlefs, N., & Frommberger, L. (2013). Machine learning for interactive systems and robots: A brief introduction. In *Proceedings of the 2nd Workshop on Machine Learning for Interactive Systems: Bridging the Gap Between Perception, Action and Communication* (pp. 19–28). New York: ACM.

Dai, P., & Goldsmith, J. (2007). Topological value iteration algorithm for Markov decision processes. In *Proceedings of the 22nd International Joint Conference on Artificial Intelligence (IJCAI'07)*, Hyderabad.

Darmois, G. (1935). Sur les lois de probabilité à estimation exhaustive. *Comptes Rendus de l'Acad'emie des Sciences Paris, 260*, 1265–1266.

Daud, A., Li, J., Zhou, L., & Muhammad, F. (2010). Knowledge discovery through directed probabilistic topic models: A survey. *Frontiers of Computer Science in China, 4*(2), 280–301.

Dempster, A., Laird, N., & Rubin, D. (1977). Maximum likelihood from incomplete data via the EM algorithm. *Journal of the Royal Statistical Society. Series B (Methodological), 39*, 1–38.

Dibangoye, J. S., Shani, G., Chaib-draa, B., & Mouaddib, A. (2009). Topological order planner for POMDPs. In *Proceedings of the 23rd International Joint Conference on Artificial Intelligence (IJCAI'09)*, Pasadena, CA.

Doshi, F., & Roy, N. (2007). Efficient model learning for dialog management. In *Proceedings of the 2nd ACM SIGCHI/SIGART Conference on Human-Robot Interaction (HRI'07)*, Arlington, VA.

Doshi, F., & Roy, N. (2008). Spoken language interaction with model uncertainty: An adaptive human-robot interaction system. *Connection Science, 20*(4), 299–318.

Doshi-Velez, F., Pineau, J., & Roy, N. (2012). Reinforcement learning with limited reinforcement: Using bayes risk for active learning in pomdps. *Artificial Intelligence, 187*. Elesiver, 115–132

Eckert, W., Levin, E., & Pieraccini, R. (1997). User modeling for spoken dialogue system evaluation. In *Proceedings of IEEE Workshop on Automatic Speech Recognition and Understanding (ASRU'97)*, Santa Barbara, CA (pp. 80–87).

Fisher, R. (1922). On the mathematical foundations of theoretical statistics. *Philosophical Transactions of the Royal Society of London. Series A, Containing Papers of a Mathematical or Physical Character, 222*(594–604), 309–368.

Fox, E. B. (2009). *Bayesian nonparametric learning of complex dynamical phenomena*. Ph.D. thesis, Massachusetts Institute of Technology.

Frampton, M., & Lemon, O. (2009). Recent research advances in reinforcement learning in spoken dialogue systems. *Knowledge Engineering Review, 24*(4), 375–408.

Gašić, M. (2011). *Statistical dialogue modelling*. Ph.D. thesis, Department of Engineering, University of Cambridge.

Gašić, M., Keizer, S., Mairesse, F., Schatzmann, J., Thomson, B., Yu, K., et al. (2008). Training and evaluation of the HIS POMDP dialogue system in noise. In *Proceedings of the 9th SIGdial Workshop on Discourse and Dialogue (SIGdial'08)*, Columbus, OH.

Georgila, K., Henderson, J., & Lemon, O. (2005). Learning user simulations for information state update dialogue systems. In *Proceedings of the 6th Annual Conference of the International Speech Communication Association (INTERSPEECH'05)*, Lisbon.

Georgila, K., Henderson, J., & Lemon, O. (2006). User simulation for spoken dialogue systems: Learning and evaluation. In *Proceedings of the 7th Annual Conference of the International Speech Communication Association (INTERSPEECH'06)*, Pittsburgh, PA.

Griffiths, T., & Steyvers, J. (2004). Finding scientific topics. *Proceedings of the National Academy of Science, 101*, 5228–5235.

Gruber, A., & Popat, A. (2007). Notes regarding computations in open htmm. http://openhtmm.googlecode.com/files/htmm_computations.pdf

Gruber, A., Rosen-Zvi, M., & Weiss, Y. (2007). Hidden topic Markov models. In *Artificial Intelligence and Statistics (AISTATS'07)*, San Juan, PR.

Hauskrecht, M. (2000). Value-function approximations for partially observable Markov decision processes. *Journal of Artificial Intelligence Research, 13*, 33–94.

Hazewinkel, M. (Ed.). (2002). *Encyclopedia of mathematics*. Berlin: Springer.

Hofmann, T. (1999). Probabilistic latent semantic analysis. In *Proceedings of the 15th Conference on Uncertainty in Artificial Intelligence (UAI'99)*, Stockholm.

Huang, J. (2005). *Maximum likelihood estimation of Dirichlet distribution parameters*. CMU Technique Report.

Ji, S., Parr, R., Li, H., Liao, X., & Carin, L. (2007). Point-based policy iteration. In *Proceedings of the 22nd National Conference on Artificial Intelligence - Volume 2 (AAAI'07)*, Vancouver, BC.

Jurafsky, D., & Martin, J. H. (2009). *Speech and language processing* (2nd ed.). Upper Saddle River, NJ: Prentice-Hall.

Kaelbling, L., Littman, M., & Cassandra, A. (1998). Planning and acting in partially observable stochastic domains. *Artificial Intelligence, 101*(1–2), 99–134.

Keizer, S., Gašić, M., Jurčíček, F., Mairesse, F., Thomson, B., Yu, K., et al. (2010). Parameter estimation for agenda-based user simulation. In *Proceedings of the 11th Annual Meeting of the Special Interest Group on Discourse and Dialogue* (pp. 116–123). Tokyo, Japan: Association for Computational Linguistics.

Kim, D., Kim, J., & Kim, K. (2011). Robust performance evaluation of POMDP-based dialogue systems. *IEEE Transactions on Audio, Speech, and Language Processing, 19*(4), 1029–1040.

Kim, D., Sim, H. S., Kim, K.-E., Kim, J. H., Kim, H., & Sung, J. W. (2008). Effects of user modeling on POMDP-based dialogue systems. In *Proceedings of the 9th Annual Conference of the International Speech Communication Association (INTERSPEECH'08)*, Brisbane.

Ko, Y., & Seo, J. (2004). Learning with unlabeled data for text categorization using bootstrapping and feature projection techniques. In *Proceedings of the 42nd Annual Meeting on Association for Computational Linguistics (ACL'04)*, Barcelona.

Koopman, B. O. (1936). On distributions admitting a sufficient statistic. *Transactions of the American Mathematical Society, 39*, 399–409.

Kotz, S., Johnson, N., & Balakrishnan, N. (2000). *Continuous multivariate distributions: Models and applications* (Vol. 1). New York: Wiley-Interscience.

Lagoudakis, M., & Parr, R. (2003). Least-squares policy iteration. *The Journal of Machine Learning Research, 4*, 1107–1149.

Lee, D., & Seung, H. (2001). Algorithms for non-negative matrix factorization. *Advances in Neural Information Processing Systems, 13*, 556–562.

Levin, E., & Pieraccini, R. (1997). A stochastic model of computer-human interaction for learning dialogue strategies. In *Proceedings of 5th European Conference on Speech Communication and Technology (Eurospeech'97)*, Rhodes.

Li, X., Cheung, W., Liu, J., & Wu, Z. (2007). A novel orthogonal nmf-based belief compression for POMDPs. In *Proceedings of the 24th International Conference on Machine learning (ICML'07)*, Corvallis.

Lison, P. (2013). Model-based bayesian reinforcement learning for dialogue management. In *Proceedings of 14th Annual Conference of the International Speech Communication Association (INTERSPEECH'13)*, Lyon.

Lusena, C., Goldsmith, J., & Mundhenk, M. (2001). Nonapproximability results for partially observable Markov decision processes. *Journal of Artificial Intelligence Research, 14*, 83–103.

Madani, O., Hanks, S., & Condon, A. (1999). On the undecidability of probabilistic planning and infinite-horizon partially observable Markov decision problems. In *Proceedings of the 16th National Conference on Artificial Intelligence (AAAI'99) and the 11th Innovative Applications of Artificial Intelligence Conference Innovative Applications of Artificial Intelligence*, Orlando, FL.

Manning, C. D., & Schütze, H. (1999). *Foundations of statistical natural language processing*. Cambridge, MA: MIT Press.

Matsubara, S., Kimura, S., Kawaguchi, N., Yamaguchi, Y., & Inagaki, Y. (2002). Example-based speech intention understanding and its application to in-car spoken dialogue system. In *Proceedings of the 19th International Conference on Computational linguistics - Volume 1*, Taipei.

Monahan, G. (1982). A survey of partially observable Markov decision processes: Theory, models, and algorithms. *Management Science, 28*, 1–16.

Neapolitan, R. (2004). *Learning Bayesian networks*. Upper Saddle River, NJ: Pearson Prentice Hall.

Neapolitan, R. (2009). *Probabilistic methods for bionformatics: With an introduction to Bayesian networks*. New York: Morgan Kaufmann.

Neu, G., & Szepesvári, C. (2007). Apprenticeship learning using inverse reinforcement learning and gradient methods. In *Proceedings of the 23rd Conference on Uncertainty in Artificial Intelligence (UAI'07)*, Vancouver, BC.

Ng, A. Y., & Russell, S. J. (2000). Algorithms for inverse reinforcement learning. In *Proceedings of the 17th International Conference on Machine Learning (ICML'00)*, Stanford, CA.

Ortiz, L. E., & Kaelbling, L. P. (1999). Accelerating EM: An empirical study. In *Proceedings of the 15th Conference on Uncertainty in Artificial Intelligence (UAI'99)*, Stockholm.

Paek, T., & Pieraccini, R. (2008). Automating spoken dialogue management design using machine learning: An industry perspective. *Speech Communication, 50*(8), 716–729.

Papadimitriou, C., & Tsitsiklis, J. (1987). The complexity of Markov decision process. *Mathematics of Operations Research, 12*(3), 441–450.

Paquet, S. (2006). *Distributed decision-making and task coordination in dynamic, uncertain and real-time multiagent environments*. Ph.D. thesis, Université Laval.

Paquet, S., Tobin, L., & Chaib-draa, B. (2005). An online POMDP algorithm for complex multiagent environments. In *Proceedings of the 4th International Joint Conference on Autonomous Agents and Multi Agent Systems (AAMAS'05)*, Utrecht.

Pieraccini, R., Levin, E., & Eckert, W. (1997). Learning dialogue strategies within Markov decision process framework. In *Proceedings of IEEE Workshop Automatic Speech Recognition and Understanding (ASRU'97)*, Rhodes.
Pietquin, O. (2004). *A framework for unsupervised learning of dialogue strategies*. Ph.D. thesis, Faculté Polytechnique de Mons.
Pietquin, O. (2006). Consistent goal-directed user model for realistic man-machine task-oriented spoken dialogue simulation. In *Proceedings of IEEE International Conference on Multimedia and Expo (ICME'06)*, Toronto, ON (pp. 425–428).
Pietquin, O., & Dutoit, T. (2006). A probabilistic framework for dialog simulation and optimal strategy learning. *IEEE Transactions on Audio, Speech, and Language Processing, 14*(2), 589–599.
Pineau, J. (2004). *Tractable planning under uncertainty: Exploiting structure*. Ph.D. thesis, Rutgers University.
Pineau, J., Gordon, G., & Thrun, S. (2003). Point-based value iteration: An anytime algorithm for POMDPs. In *International Joint Conference on Artificial Intelligence (IJCAI'03)*, Acapulco.
Pineau, J., West, R., Atrash, A., Villemure, J., & Routhier, F. (2011). On the feasibility of using a standardized test for evaluating a speech-controlled smart wheelchair. *International Journal of Intelligent Control and Systems, 16*(2), 124–131.
Pitman, E. (1936). Sufficient statistics and intrinsic accuracy. *Proceedings of the Cambridge Philosophical Society, 32*, 567–579.
Png, S., & Pineau, J. (2011). Bayesian reinforcement learning for POMDP-based dialogue systems. In *Proceedings of the IEEE International Conference on Acoustics, Speech, and Signal Processing (ICASSP'11)*, Prague.
Png, S., Pineau, J., & Chaib-Draa, B. (2012). Building adaptive dialogue systems via bayes-adaptive pomdps. *IEEE Journal of Selected Topics in Signal Processing, 6*(8), 917–927.
Poupart, P., & Boutilier, C. (2002). Value-directed compression of POMDPs. In *Advances in Neural Information Processing Systems 14 (NIPS'02)*, Vancouver, BC.
Rabiner, L. R. (1990). A tutorial on hidden Markov models and selected applications in speech recognition. In *Readings in speech recognition* (pp. 267–296). San Francisco: Morgan Kaufmann Publishers.
Ramachandran, D., & Amir, E. (2007). Bayesian inverse reinforcement learning. In *Proceedings of the 20th International Joint Conference on Artificial Intelligence (IJCAI'07)*, Hyderabad.
Rieser, V., & Lemon, O. (2006). Cluster-based user simulations for learning dialogue strategies. In *Proceedings of the 7th Annual Conference of the International Speech Communication Association (INTERSPEECH'06)*, Pittsburgh, PA.
Rieser, V., & Lemon, O. (2011). Reinforcement learning for adaptive dialogue systems: a data-driven methodology for dialogue management and natural language generation. Springer Science & Business Media.
Robert, C. P., & Casella, G. (2005). *Monte Carlo statistical methods. Springer texts in statistics*. Secaucus, New York: Springer.
Ross, S., Chaib-draa, B., & Pineau, J. (2007). Bayes-adaptive POMDPs. In *Proceedings of the 21st Annual Conference on Neural Information Processing Systems (NIPS'07)*, Vancouver, BC.
Ross, S., Pineau, J., Chaib-draa, B., & Kreitmann, P. (2011). A Bayesian approach for learning and planning in partially observable Markov decision processes. *Journal of Machine Learning Research, 12*, 1729–1770.
Ross, S., Pineau, J., Paquet, S., & Chaib-draa, B. (2008). Online planning algorithms for POMDPs. *Artificial Intelligence Research, 32*(1), 663–704.
Roy, N., Gordon, J., & Thrun, S. (2005). Finding approximate POMDP solutions through belief compression. *Journal of Artificial Intelligence Research, 23*, 1–40.
Roy, N., Pineau, J., & Thrun, S. (2000). Spoken dialogue management using probabilistic reasoning. In *Proceedings of the 38th Annual Meeting on Association for Computational Linguistics (ACL'00)*, Hong Kong.
Russell, S., & Norvig, P. (2010). *Artificial intelligence: A modern approach*. New York: Prentice Hall.

Schatzmann, J., Thomson, B., Weilhammer, K., Ye, H., & Young, S. (2007). Agenda-based user simulation for bootstrapping a POMDP dialogue system. In *Human Language Technologies 2007: The Conference of the North American Chapter of the Association for Computational Linguistics; Companion Volume, Short Papers* (pp. 149–152). Association for Computational Linguistics.

Schatzmann, J., Weilhammer, K., Stuttle, M., & Young, S. (2006). A survey of statistical user simulation techniques for reinforcement-learning of dialogue management strategies. *Knowledge Engineering Review, 21*(2), 97–126.

Schatzmann, J., & Young, S. (2009). The hidden agenda user simulation model. *IEEE Transactions on Audio, Speech, and Language Processing, 17*(4), 733–747.

Scheffler, K., & Young, S. (2000). Probabilistic simulation of human-machine dialogues. In *Proceedings of IEEE International Conference on Acoustics, Speech, and Signal Processing (ICASSP'00)* (Vol. 2, pp. 1217–1220).

Smallwood, R., & Sondik, E. (1973). The optimal control of partially observable Markov processes over a finite horizon. *Operations Research, 21*, 1071–1088.

Smith, T., & Simmons, R. (2004). Heuristic search value iteration for pomdps. In *Proceedings of the 20th Conference on Uncertainty in Artificial Intelligence (UAI '04)*, Banff, AB.

Sondik, E. (1971). *The optimal control of partially observable Markov processes*. Ph.D. thesis, Stanford University.

Spaan, M., & Spaan, N. (2004). A point-based POMDP algorithm for robot planning. In *Proceedings of IEEE International Conference on Robotics and Automation (ICRA'04)*, New Orleans, LA.

Spaan, M., & Vlassis, N. (2005). Perseus: Randomized point-based value iteration for POMDPs. *Journal of Artificial Intelligence Research, 24*(1), 195–220.

Sudderth, E. B. (2006). *Graphical models for visual object recognition and tracking*. Ph.D. thesis, Massachusetts Institute of Technology.

Sutton, R. S., & Barto, A. G. (1998). *Reinforcement learning: An introduction*. Cambridge: MIT Press.

Syed, U., & Schapire, R. (2008). A game-theoretic approach to apprenticeship learning. In *Proceedings of the Twenty-First Annual Conference on Neural Information Processing Systems*, Vancouver, BC.

Thomson, B. (2009). *Statistical methods for spoken dialogue management*. Ph.D. thesis, Department of Engineering, University of Cambridge.

Thomson, B., & Young, S. (2010). Bayesian update of dialogue state: A POMDP framework for spoken dialogue systems. *Computer Speech and Language, 24*(4), 562–588.

Traum, D. (1994). *A computational theory of grounding in natural language conversation*. Ph.D. thesis, University of Rochester.

Watkins, C. J. C. H., & Dayan, P. (1992). Technical note Q-Learning. *Machine Learning, 8*, 279–292.

Weilhammer, K., Williams, J. D., & Young, S. (2004). The SACTI-2 corpus: Guide for research users. Cambridge University. Technical Report.

Welch, L. (2003). Hidden Markov models and the Baum-Welch algorithm. *IEEE Information Theory Society Newsletter, 53*(4), 1–10.

Wierstra, D., & Wiering, M. (2004). Utile distinction hidden Markov models. In *Proceedings of the Twenty-First International Conference on Machine Learning* (p. 108). New York: ACM.

Williams, J. D. (2006). *Partially observable Markov decision processes for spoken dialogue management*. Ph.D. thesis, Department of Engineering, University of Cambridge.

Williams, J. D., & Young, S. (2005). The SACTI-1 corpus: Guide for research users. Department of Engineering, University of Cambridge. Technical Report.

Williams, J. D., & Young, S. (2007). Partially observable Markov decision processes for spoken dialog systems. *Computer Speech and Language, 21*, 393–422.

Young, S., Gasic, M., Thomson, B., & Williams, J. D. (2013). Pomdp-based statistical spoken dialog systems: A review. *Proceedings of the IEEE, 101*(5), 1160–1179.

Zhang, B., Cai, Q., Mao, J., Chang, E., & Guo, B. (2001a). Spoken dialogue management as planning and acting under uncertainty. In *Proceedings of the 9th European Conference on Speech Communication and Technology (Eurospeech'01)*, Aalborg.
Zhang, B., Cai, Q., Mao, J., & Guo, B. (2001b). Planning and acting under uncertainty: A new model for spoken dialogue system. In *Proceedings of the 17th Conference in Uncertainty in Artificial Intelligence (UAI'01)*, Seattle, Washington.
Ziebart, B., Maas, A., Bagnell, J., & Dey, A. (2008). Maximum entropy inverse reinforcement learning. In *Proceedings of the 23rd National Conference on Artificial Intelligence (AAAI'08)*, Chicago, IL.

Printed in the United States
By Bookmasters